LEARNING AND ASSESSING SCIENCE PROCESS SKILLS

THIRD EDITION

RICHARD J. REZBA
Virginia Commonwealth University

CONSTANCE STEWART SPRAGUE
Indiana University South Bend

RONALD L. FIEL
Morehead State University

H. JAMES FUNK

JAMES R. OKEY
University of Georgia

HAROLD H. JAUS
University of North Carolina at Charlotte

 KENDALL/HUNT PUBLISHING COMPANY
4050 Westmark Drive   P.O. Box 1840   Dubuque, Iowa 52004-1840

Disclaimer

Adult supervision is required when working on projects. Use proper equipment (gloves, forceps, safety glasses, etc.) and take other safety precautions such as tying up loose hair and clothing and washing your hands when the work is done. Use extra care with chemicals, dry ice, boiling water, or any heating elements. Hazardous chemicals and live cultures (organisms) must be handled and disposed of according to appropriate directions from your adult advisor. Follow your science fair's rules and regulations and the standard scientific practices and procedures required by your school. No responsibility is implied or taken for anyone who sustains injuries as a result of using the materials or ideas, or performing the procedures described in this book.

Additional safety precautions and warnings are mentioned throughout the text. If you use common sense and make safety a first consideration, you will create a safe, fun, educational, and rewarding project.

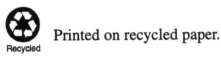

Printed on recycled paper.

Recycled

Student activity pages and tables that are clearly designated
with: Rezba, Sprague, Fiel, Funk, Okey, & Jaus,
LEARNING AND ASSESSING SCIENCE PROCESS SKILLS,
©Kendall/Hunt 1995, are the only pages in this book that may be
reproduced without written permission.

Previously entitled *Learning Science Process Skills*

Copyright 1979, 1985, 1995 by Kendall/Hunt Publishing Company

ISBN 0-8403-8430-0

Printed in the United States of America

10 9 8 7 6 5

DEDICATION

The authors dedicate this edition to H. James Funk who is remembered for his humor, his love of science, and his concern for our nation's children.

The goal setting activity on page xviii is in keeping with Dr. Funk's high expectations for those individuals whose chosen career is to teach children.

iv

CONTENTS

PREFACE

Teaching science is an awesome responsibility. Regardless of how you personally view science, the children you teach are depending on you to model good science and to teach them the skills they need to learn more about our increasingly scientific and technological world. *Learning and Assessing Science Process Skills* is designed to help you develop the knowledge and skills necessary to bring the science process skills to your students.

What are the science process skills? They are the things that scientists do when they study and investigate. Observing, measuring, inferring, and experimenting are among the thinking skills used by scientists or by you and your students when doing science. Much of the pleasure of both learning and teaching science is **experiencing** science. Mastering these process skills will help you develop the kind of science program that mirrors real science.

ORGANIZATION

Learning and Assessing Science Process Skills is presented in two parts. In Part One you will learn and practice the skills of observing, communicating, classifying, measuring, inferring, and predicting. These skills are called The Basic Science Process Skills because they form the foundation for later and more complex thinking skills. Instruction on the Basic Science Process Skills begins in pre-school, is emphasized in the elementary grades, and continues into middle school and beyond. In Part Two you will learn the skills you and your students need to design and conduct scientific investigations in class and at home. These skills are known as The Integrated Science Process Skills because they are used together to do what many consider the ultimate in problem solving in science - *experimenting*.

Although Parts One and Two are primarily about helping you become *competent* in the science process skills, they are also about helping you become *confident* in your ability to help students learn the same skills. While you are learning these skills, you will also be learning *how* you learned them. In sections called, **Decision Making 1 and 2**, you will be asked to use what you learned to make instructional decisions about how process skill development can be enhanced in existing curriculum materials.

NEW FEATURES OF THE THIRD EDITION

Although originally intended as a text for K - 8 teachers to learn the skills necessary to develop a hands-on and minds-on science program, earlier editions soon served other needs as well. For more than a decade, this book has been a source of both instructional and assessment ideas for practicing classroom teachers and curriculum developers. In this new edition, assessment strategies are emphasized from beginning to end of the book. Assessment items are provided as self-checks throughout the text and in each chapter's Self-Assessment. A new feature, **Student Assessment**, is now included at the end of each chapter as an example of how student mastery of that process skill might be assessed. The 16 **Student Assessment** examples represent nine different strategies for assessing the science process skills - from high tech optically scanned teacher checklists to authentic performance tasks.

Teaching is about decision making. An already overcrowded curriculum prevents the easy decision of just adding a new unit on measuring skills or another on identifying variables. A more important decision making skill involves the ability to recognize opportunities in existing curriculum to enhance the teaching, learning, and assessing of the science process skills. In Decision Making 1 and 2, examples of typical science activities have been added to Parts One and Two to help you learn how to modify instructional materials to help students learn new skills. In addition, new instructional activities as well as continuing favorites can be easily adapted for classroom use. Helpful ways to encourage young children to use the basic process skills are now included as **Thought Starters** in Part One. Finally, a fresh new look to *Learning and Assessing Science Process Skills* not only is more user-friendly, but represents more than ever the timely significance of learning the science process skills.

A CHALLENGE

With every adventure there lies a risk. The risk here is that the individual chapters of the text imply a separation of the process skills used to do science. In truth these thinking skills of science are interdependent. As you study the science process skills, we hope the artificial separations created here will dissolve and by merging these skills you become more able and confident in providing exemplary science instruction for your students.

The Authors

INTRODUCTION

Reform Initiatives for Teaching and Learning Science

The Industrial Age that helped make America great relied on our country's natural resources and the hard work of our people. Although a strong work ethic and the wise use of our natural resources continue to be important in the Information Age, *working smarter not just harder* will secure our position in an increasingly global marketplace. Major changes in our society and in the world are causing a rethinking of the purposes of education, especially in mathematics and science. The goals for science education in the nineties and beyond stress science as ways of thinking and investigating as well as a body of knowledge.

Ways of thinking in science are called the process skills. When scientists and students *do* science they are using such thinking skills as inferring, classifying, hypothesizing, and experimenting. The science process skills, along with the knowledge those skills produce, and the scientific values and habits of mind define the nature of science. Unfortunately, the teaching and learning of science does not always reflect the true nature of science. Too often students are burdened with short–lived learning of facts and boldfaced terminology at the expense of ever actively doing science. An increasing body of research supports the notion that students learn best when actively engaged, both physically and mentally, in hands–on and minds–on activities. There is also growing agreement among teachers and policy makers that *less is more*, where focusing greater attention on fewer concepts and skills is far more beneficial to students than covering vast amounts of abstract science content. Hands–on activities leads to minds–on understanding when teachers can concentrate on a more manageable number of big ideas and where students have regular opportunities to *think* about what they have been doing.

The *National Science Education Standards Project* sponsored by the National Academy of Science, *Project 2061* developed by the American Association for the Advancement of Science (AAAS), and the *Scope, Sequence, and Coordination Project* of the National Science Teachers Association (NSTA) are all long term reform initiatives in science education that will extend into the next century. The standards, benchmarks, and guidelines of these national reform efforts uniformly emphasize the need to create learning environments that encourage students' understanding of the scientific endeavor as well as students' excitement and enjoyment in its pursuit. High standards characterize these current reform initiatives because low expectations for students leads to fact–driven memorization with few opportunities to explore and to experiment using the science process skills.

Unlike the curricular reforms that followed the launching of the Russian Sputnik in 1957, there is a genuine commitment to science for all students, not just the elite. Earlier reforms were largely aimed at guaranteeing an adequate supply of scientists and engineers to meet our national needs. Current reform efforts reject practices where populations of students defined by gender, race, ethnicity, physical disability, and economic status are excluded from opportunities to learn science and are discouraged from pursuing science as a career. There is general consensus that science belongs to everyone and that it is in our nation's best interest that all students become scientifically literate.

Assessment in the Service of Reform

> *By the end of the following grades, students should*
>
> **2nd grade:** *Raise questions about the world around them and be willing to seek answers to some of them by making careful observations and trying things out.*
>
> **5th grade:** *Offer reasons for their findings and consider reasons suggested by others.*
>
> **8th grade:** *Know that hypotheses are valuable, even if they turn out not to be true, if they lead to fruitful investigations.*
>
> <u>Benchmarks for Science Literacy</u>, AAAS' Project 2061

There is something contradictory about teaching students to carefully observe objects and then testing them with, "Which of the following is a definition of quantitative observation?". Knowing a definition may be important, but it is never sufficient. It is often said that assessment drives instruction. But what if science education is being driven where you do not want to go? Clearly, a conflict emerges when reform–based efforts in science are moving in one direction while textbook exams and standardized tests are staying the traditional course.

New directions in assessment suggest a breaking down of distinctions between assessment and instruction and for a testing program that includes assessments that are more *authentic*, that is, more representative of real–world activities. The movement toward more performance–based assessment is consistent with the directions proposed by the national reform efforts in science, such as Project 2061's benchmarks. Benchmarks are goals that serve as the core of the knowledge, skills, and habits of mind that contribute to the scientific literacy of all students.

Assessment is the process of gathering evidence of students' abilities and achievements. It is a form of communication that provides the information so evaluations can be made. Assessments help teachers and parents determine what students know, are able to do, and what they still need to learn. Evidence about student performance in science is needed for a variety of purposes, such as making instructional decisions, tracking student progress, communicating judgements, and evaluating programs. For each of these purposes, data are collected from essentially three basic sources: observations, student responses to questions, and student products and performances. Science assessment should be about gathering evidence of students' achievements and then using that evidence to further the growth of science knowledge and skills for all students.

Assessment methods should allow students to demonstrate what they know and are able to do, not just what they do not know. Consistent with this view is a growing movement away from the almost exclusive use of multiple choice assessment to *multiple forms of assessment*, including the multiple choice format. While multiple choice and other forced choice formats will remain important assessment tools for teachers, new more thought–provoking forms of assessment are needed. Just as lecturing should not be the sole method of instruction, neither should multiple choice tests be the only means of assessment. Multiple forms of assessment include open–response questions, performance tasks, portfolios, interviews, teacher rating forms, and a variety of checklists for teachers, students, parents, and peers.

One characteristic of other forms of assessment, sometimes called alternative assessment, is their ability to connect assessment with instruction. Most current testing practices are only assessment tools, not teaching tools. Assessment occurs only when instruction stops. It is interesting to note that assessment comes from the French *assidere* (Latin *sedere*) meaning to *sit beside*, suggesting that a much closer relationship should exist between instruction and assessment. A variety of innovative methods is needed to bridge the gap between teaching and assessment. Assessments in science should allow students to use their science process skills and content knowledge from the science disciplines in much the same way as they do in science class. Assessment should mirror the science that is most important for students to learn.

Multiple forms of assessment are also consistent with what we know about learning. Students with differing learning styles should have varied opportunities to demonstrate what they know and are able to do in science. Assessment needs to facilitate each student's continued learning in science.

The changing world and the reform efforts in science instruction require new assessments. If current assessment practices continue, reform in science education is not likely to occur. Fortunately, there are superb teachers who systematically observe, challenge, and listen to students to lead the way in assessing the performance of students. Their exemplary science classrooms are characterized by high expectations, challenging tasks, strong work ethic, mutual respect, and a belief in science for all students. Teachers are the richest sources of information about students. They have always been and will continue to be the major assessors of student achievement. Of the multiple ways to assess students, nine forms of assessment are described here and modeled in the 16 chapters of *Learning and Assessing Science Process Skills* (see Figure 1, Multiple Forms of Assessment on the next page).

FIGURE 1. MULTIPLE FORMS OF ASSESSMENT

ASSESSMENT	TOPIC	CHAPTER
Open Response Question	Observing Keeping Factors Constant	1 12
Performance Task	Classifying Measuring Inferring Predicting Relationships Between Variables Defining Variables Experimenting	3 4 5 6 10 14 16
Portfolio	Identifying Variables	7
Teacher Paper and Pencil Checklist	Communicating	2
Self/Peer/Family Checklist	Tabulating Data	8
Teacher Optical Scanned Checklist	Laboratory Behaviors	11
Teacher Rating Sheet	Graphing Skills	9
Individual Performance Within a Group Rating Form	Designing Investigations	15
Interview	Constructing Hypotheses	13

OPEN–RESPONSE QUESTIONS

Open–response (also called open–ended) questions provide students with the opportunity to make observations, analyze investigations, solve problems, and design experiments by constructing their own responses in writing or by drawing. A range of possible responses is typical because students are asked to construct a response rather than choose from a presented set. Questions typically contain two parts, the *Directions to the Student* and the *Prompt*. The directions tell students what is expected of them, while the prompt provides the scenario and necessary information for the problem. Scoring guidelines, often called **rubrics**, are developed specific to each open–response question because these problems have a variety of correct responses. Although rubrics can take many forms, they often consist of a scale with 4, 5, or 6 levels of proficiency (see **Student Assessment** in Chapter 12). At the highest level a student demonstrates an in–depth understanding of the concept and an ability to communicate it effectively, while the reverse is true at the lowest level where little or no evidence of concept mastery or ability to communicate ideas is evidenced. Examples of open–response questions are provided in Chapters 1 and 12.

PERFORMANCE TASKS

Performance tasks in science are activities in which students can demonstrate their knowledge and higher order thinking skills by manipulating equipment and materials and recording their observations and conclusions. They provide students with the opportunity to demonstrate their understanding of important scientific processes and concepts by actually doing science as part of the test. Manipulating materials is characteristically part of the assessment. Performance tasks can be completed individually or by groups. As in the case of open–response questions, detailed scoring guidelines, or rubrics, are required. Examples of performance tasks to assess both the basic and integrated process skills are given in Chapters 3, 4, 5, 6, 10, 14, and 16.

PORTFOLIOS

Portfolios may be broadly defined as a collection of representative work including some evidence that the student has evaluated the quality of his or her own work. While portfolios have long been used by artists and models to evidence their accomplishments, their use in schools has been limited until recently. A growing trend, originally in the language arts, has been to use portfolios to create a cumulative record of students' growth. Portfolios should evidence significant tasks, worthy of time and commitment. In science one such significant task is designing and conducting an experiment, beginning with the ability to identify variables. Portfolio assessment allows teachers to track students' progress toward high standards by collecting evidence of students' ability to identify variables, construct hypotheses, tabulate and graph data, write conclusions, and so on.

Portfolio assessment is often more time consuming and more difficult to manage than other forms of assessment, but the variety of benefits is also significant, such as student growth in self–assessment and confidence. The science research checklist and portfolio evaluation form provided in the **Student Assessment** example in Chapter 7 helps students learn to monitor and evaluate their own work.

CHECKLISTS

Several forms of checklists for assessing the science process skills are illustrated in Chapters 2, 8, and 11:

Paper and pencil checklists were traditionally used by primary teachers to record observations of individual students. These observations provided evidence of successful mastery of numerous academic and social skills such as reading from left to right and willingness to share materials. An example using communication skills is provided in Chapter 2.

Self/Peer/Family checklists allow parents, students, and peers to take more active roles in assessing student progress. Providing students a set of criteria for evaluating their own work shifts some of the responsibility for evaluation to them. Checklists of this type as illustrated in Chapter 8 can help students learn to monitor their own work and to reflect on their performance as well as inform parents of specific expectations.

Teacher Optical Scanned Checklist Recent advances in technology are making possible new classroom tools that were once only science fiction. One such tool uses barcodes, optical scanning technology, and a computer to allow teachers to easily record, store, and report observations of individual students. While initial reaction by parents and others may be a concern about high tech depersonalization, this technology as well as other emerging tools (such as computer interactive note pads) allows teachers to provide a level of individual attention never before possible. Note in Chapter 11 how students' laboratory behaviors – following safety procedures, caring for materials, and the active use of the process skills – can be easily observed and recorded using this new technology.

TEACHER RATING SHEETS

When desired student outcomes consist of several steps or sub–skills, teacher rating sheets serve the dual role of focusing students' attention on these steps as well as a tool for evaluating the final product. Teacher rating sheets can be used for any of the process skills that consist of a number of sub–skills, such as for graphing in Chapter 9.

GROUP RATING FORM

In business, industry, scientific research, and other aspects of real life, knowledge is frequently developed by groups rather than by individuals working alone. Team work and cooperation are increasingly common. Similarly, the growing use of cooperative learning at school demands assessments that measure the value of group interaction, yet fairly appraise an individual's contribution to group accomplishments. The group rating form illustrated in Chapter 15 is one attempt at meeting this challenging task.

INTERVIEWS

Few things in life are free. The positive features of multiple forms of assessment often come at a price – time, planning, and thought on the part of both teachers and students. When student interviews can be managed, there is great potential for gaining insight into students' conceptions as well as their misconceptions because of the interaction between teacher and student. An example of the interview technique is provided in Chapter 13 and is taken from *Science – A Process Approach II*, originally developed by the American Association for the Advancement of Science.

HOW TO USE THIS BOOK

Much of what you learn from this book depends on how you use it. Begin each chapter by first reading the purpose and objectives; knowing what is expected of you will help you learn the skills presented in the chapter. Do all of the activities because mastering the science process skills can only be achieved by being actively involved. Most activities have self–check sections that provide you with immediate feedback on your responses. The special symbol, ✓ will cue you that the answers to the self–check questions will appear next. Write your responses before reading the answers. The eye is sometimes quicker than the conscience so use a sheet of paper to cover the answers and slide the paper down the page as you proceed. At the end of each chapter a Self–Assessment is provided for you to demonstrate your knowledge and skills. Use the answers provided to check your level of mastery. Experience with the chapter self–checks, self–assessments, and student assessment examples will also help you consider how you will assess your own students' abilities to use the science process skills. As you learn the process skills, think about how that experience will help you decide how to teach these same skills to your students. Study the examples in the sections called **Decision Making Skills 1 and 2** to help you enhance the teaching of process skills.

The activities in the book were designed for either individual or small group study. You are encouraged, however, to work cooperatively with at least one other person as you proceed through the chapters. Working together may help you better process information, practice skills, receive feedback, and have more fun learning.

If you are using this book in a course, your instructor may suggest different materials to use in class or at home as you complete the activities in this book. Using different materials may result in some of your answers being somewhat different than ours. Those of you who are in–service teachers using ideas and activities from this book with your students may wish to substitute one kind of material for another, such as plastic cups for baby food jars. In addition, you may wish to make overhead transparencies of some pages for use in your instruction and to use assessment examples throughout the book as part of your assessment program.

References

Center for the Study of Testing, Evaluation, and Educational Policy (1992). *The Influence of Testing on Teaching Math and Science in Grades 4–12.* Boston: Boston College.

Champagne, A., Lovitts, B., and Calinger, B. (1990). *Assessment in the Service of Instruction.* Washington, DC: American Association for the Advancement of Science.

Cothron, J., Giese, R., and Rezba, R. (1993). *Students and Research: Practical Strategies for Science Classrooms and Competitions, 2nd ed.* Dubuque, Iowa: Kendall/Hunt Publishing Company.

Hart, D. (1994). Authentic Assessment: *A Handbook for Educators.* New York: Addison–Wesley Publishing Company.

Kulm, G. and Malcom, S., eds. (1991). *Science Assessment in the Service of Reform.* Washington, DC: American Association for the Advancement of Science.

National Committee on Science Education Standards and Assessment (1993). *National Science Education Standards: July '93 Progress Report.* Washington, DC: National Research Council.

National Science Teachers Association (1992). *The Scope, Sequence and Coordination Content Core, Volume I.* Washington, DC: The National Science Teachers Association.

National Science Teachers Association (1992). *Scope, Sequence and Coordination of Secondary School Science, Relevant Research, Volume II.* Washington, DC: The National Science Teachers Association.

Project 2061, American Association for the Advancement of Science (1993). *Benchmarks for Science Literacy.* New York: Oxford University Press, Inc.

Rutherford, F. J., and Ahlgren, A. (1990). *Science for All Americans.* New York: Oxford University Press, Inc.

SETTING GOALS

Children benefit most from those teachers who set goals for themselves. Think about what you would like to achieve in your first three years of teaching science. Write the first few sentences of a newspaper article that describe what you would like written about the way you teach science to elementary and middle school students.

Note: Write the newspaper article in pencil. After you have completed this book you will be asked to return to this article to make any changes you wish that reflect changes in your goals.

The World Reporter

Volume CXVII

Teacher Achieves 3-Year Goal

PART ONE

Basic Science Process Skills

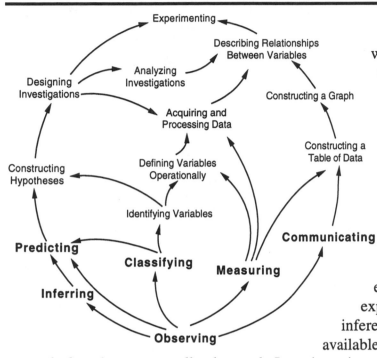

The basic science process skills are what people do when they *do* science. Children using these same skills are active learners. They use their senses to *observe* objects and events and they look for patterns in those observations. They *classify* to form new concepts by searching for similarities and differences. Orally and in writing, they *communicate* what they know and are able to do. To quantify descriptions of objects and events, they *measure*. They *infer* explanations and willingly change their inferences as new information becomes available. And they *predict* possible outcomes before they are actually observed. Learning science this way may be very different from the way you learned science in your elementary and middle school experiences. For you to teach the science process skills to children and to be able to implement a science curriculum that emphasizes these skills, you must first learn them yourself.

While learning the basic science process skills, you too will be an active learner. The activities in Part One have been carefully designed to help you focus on the process skills, and all of them have been successfully used in elementary and middle school classrooms. By thinking about how these activities helped you learn these skills, you can look for similar activities to help your students learn the process skills in much the same way.

The activities in this book use simple, ordinary supplies. Good science can be learned with everyday materials; elaborate and costly equipment is not required. In fact, it is the very ordinary that often stimulates students to ask questions that lead them to fruitful inquiry.

You will begin Part One with the skill of *observing*. This is the science process skill on which all the others are based. Each time you learn a new skill, ask yourself these two questions:

> ***How am I learning this skill?***
> ***How will I teach this skill to students?***

Teaching Children

Answering these questions will help you think about teaching as well as learning the process skills. After you have completed Part One, Basis Science Process Skills, you will be asked to make some instructional decisions about how you might teach these same skills to children.

1

Observing

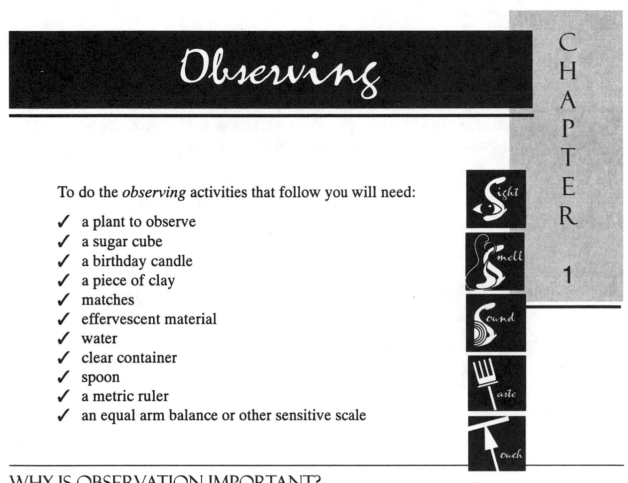

To do the *observing* activities that follow you will need:

✓ a plant to observe
✓ a sugar cube
✓ a birthday candle
✓ a piece of clay
✓ matches
✓ effervescent material
✓ water
✓ clear container
✓ spoon
✓ a metric ruler
✓ an equal arm balance or other sensitive scale

WHY IS OBSERVATION IMPORTANT?

By observing we learn about the fantastic world around us. We observe objects and natural phenomena through our five senses: sight, smell, touch, taste, and hearing.

The information we gain leads to curiosity, questions, interpretations about our environment, and further investigation. Ability to observe is the most basic skill in science and is essential to the development of other science process skills such as inferring, communicating, predicting, measuring, and classifying.

The purpose of these exercises is to help you further develop your skills of observation and learn about different kinds of observations you can make about your environment.

B. C. by permission of Johnny Hart and Field Enterprises, Inc.

PERFORMANCE OBJECTIVES

After completing this set of activities you should:

1. given an object, substance, or event, be able to construct a list of qualitative and quantitative observations about an object, substance, or event. Your observation must be perceived through at least four of your senses.
2. given an event in which a change is involved, be able to construct a list of qualitative and quantitative observations about the changes before, during, and after they occur.

To observe an object or substance means to carefully explore all of its properties. Objects may have such properties as color, texture, odor, shape, weight, volume, or temperature. They may even make sounds either on their own or when manipulated.

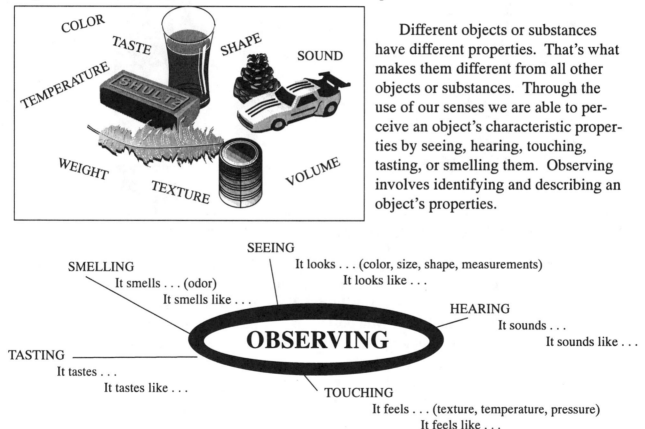

Different objects or substances have different properties. That's what makes them different from all other objects or substances. Through the use of our senses we are able to perceive an object's characteristic properties by seeing, hearing, touching, tasting, or smelling them. Observing involves identifying and describing an object's properties.

SMELLING
It smells . . . (odor)
It smells like . . .

SEEING
It looks . . . (color, size, shape, measurements)
It looks like . . .

HEARING
It sounds . . .
It sounds like . . .

OBSERVING

TASTING
It tastes . . .
It tastes like . . .

TOUCHING
It feels . . . (texture, temperature, pressure)
It feels like . . .

Activity 1

USING ALL YOUR SENSES

⇨ **GO** to one of the plants in the room, or at home, and gather as much information as you can about the plant using all your senses except taste. (CAUTION: Tasting any unknown substance is hazardous business; never taste anything unless you are absolutely certain that there is no danger involved.) Below, list at least ten observations about the plant. For each observation record the sense you used to obtain the information.

Remember to cover the answers in the self-check and refer to them only after completing the activity.

Observation	Senses
1.	
2.	
3.	
4.	
5.	
6.	
7.	
8.	
9.	
10.	

Compare your observations with someone else's or with those below.

SELF-CHECK ✓

Your list of observations should provide at least enough information to answer these questions about the plant you observed:

1. What color is it? Is the color evenly distributed? (sight)
2. Is the plant tall, short, spindly, sprawling? (sight)
3. Is there one main stem or many? (sight)
4. What is the general shape of the leaves? (sight)
5. Do the leaves have jagged or smooth edges? (sight)
6. Are the leaves shiny or dull? (sight, touch)
7. Are the leaves opposite one another or alternate? (sight)
8. Are the veins of the leaves distinct? Is there a central vein? Are the veins opposite one another or alternate? (sight)
9. Is the stem thick or thin? (sight, tough)
10. Are the leaves in clusters or separate? (sight)
11. What is the texture of the stem and leaf surfaces? (touch)
12. Do the leaves feel waxy? (touch)
13. Are the leaves stiff or easily pliable? (touch)
14. Does any part of the plant have an odor? (smell)

Activity 2

MAKING QUALITATIVE AND QUANTITATIVE OBSERVATIONS

Most of your observations in Activity 1 were probably qualitative observations; that is, you used only your senses to obtain the information. The following statements are examples of qualitative observations you might have made:

- It is light green in color. (sight)
- It has a pungent odor. (smell)
- It tastes sour. (taste)
- Its leaves are waxy and smooth. (touch)
- It makes a rustling sound when lightly rubbed. (hearing)

Sometimes we want more precise information than our senses alone can give us and we include a reference to some standard unit of measure. Observations that involve number or quantity are quantitative observations. Quantitative observations help us communicate specifics to others and provide a basis for comparisons. The following statements are examples of quantitative observations that could be made about a plant.

- One leaf is 10 cm long and 6 cm wide. (metric ruler)
- The mass of one leaf is 5g. (balance)
- The temperature of the room in which it grows is 22°C. (thermometer)
- This plant's leaves are clustered in groups of five.
- This plant is larger than that plant.
- Each flower is as wide as 3 paper clips placed end to end.

Quantitative observations made with instruments such as rulers, meter sticks, balances, and graduated cylinders or beakers, give us specific and precise information. Although approximations and comparisons are not as precise, they are also quantitative observations.

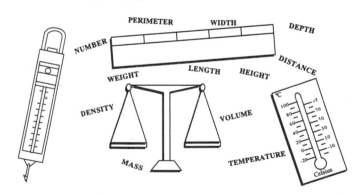

⇨ **Go** to the observation supply area and pick up one of the small white cubes. In the charts that follow, list at least five qualitative observations and four quantitative observations about this object. For each qualitative observation identify which sense you used to gain the information and for each quantitative observation identify the instrument you used to aid your senses. In this activity you may use all five senses.

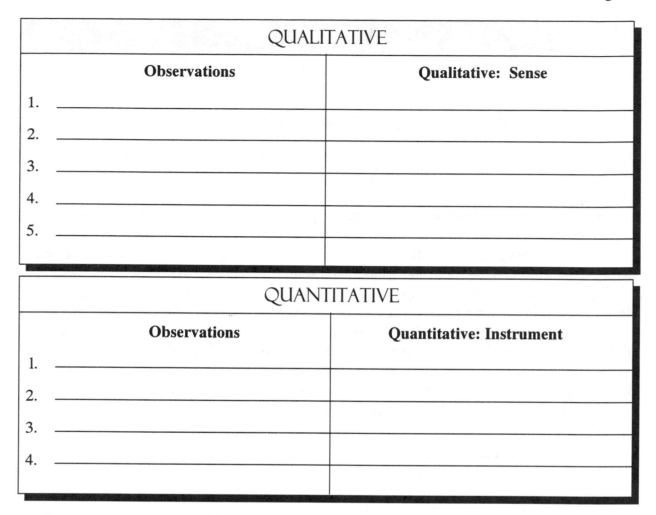

QUALITATIVE

Observations	Qualitative: Sense
1. _____	
2. _____	
3. _____	
4. _____	
5. _____	

QUANTITATIVE

Observations	Quantitative: Instrument
1. _____	
2. _____	
3. _____	
4. _____	

Compare your answers with someone else's or check your answers with those that follow.

SELF-CHECK ✓

Some of the observations you may have made are included in the following chart. Of course there are other acceptable observations. If you are in doubt about any of your observations, please ask a peer or your instructor.

Observations	Qualitative: Sense
1. Object is cube shaped, white, sparkles.	Qualitative: Sight
2. Tastes sweet.	Qualitative: Taste
3. Has no distinctive odor.	Qualitative: Smell
4. Feels hard but crumbly; rough texture.	Qualitative: Touch
5. Makes a sharp sound when dropped	Qualitative: Hearing

Observations	Quantitative: Instrument
1. Length: 1.3 cm	Quantitative: Metric ruler
2. Width: 1.3 cm	Quantitative: Metric ruler
3. Height: 1.3 cm	Quantitative: Metric ruler
4. Mass: 2.0 g	Quantitative: Balance

$\mathcal{A}$ ctivity 3

OBSERVING CHANGES

You will often observe objects or phenomena that undergo physical or chemical changes. Your observations will be either qualitative, in which you use your senses to obtain information, or quantitative, in which you make a reference to some standard unit of measure. When asked to describe a change, it is important to include statements of observation made before, during, and after the change occurs.

Think, for example, about the changes you might observe when you make popcorn. Before the kernel is heated, it is teardrop shaped, about 1 cm x 0.5 cm x 0.5 cm in size, light brown in color, and has a hard, smooth shell. During the change (popping) the shell splits, a white puffy mass expands through the shell, and a short, light sound is produced. After the change the piece of popcorn is irregular in shape, about 3 cm x 2 cm x 3 cm in size, has a white, puffy texture, and a corn like taste. Of course, more observations could be made.

⇨ *Go* to the observation supply area and pick up a birthday candle, a piece of clay for a base, a ruler, and matches. In the chart below describe the candle before, during, and after it is burned. Include in your description at least seven (4 qualitative and 3 quantitative) observations before the change, three general statements about the changes as you observe them occurring, and five observations (4 qualitative and 1 quantitative) after the change has taken place.

Qualitative Observations	Quantitative Observations
Before	
1. _____	1. _____
2. _____	2. _____
3. _____	3. _____
4. _____	
5. _____	
During	
1. _____	
2. _____	
3. _____	
After	
1. _____	1. _____
2. _____	
3. _____	

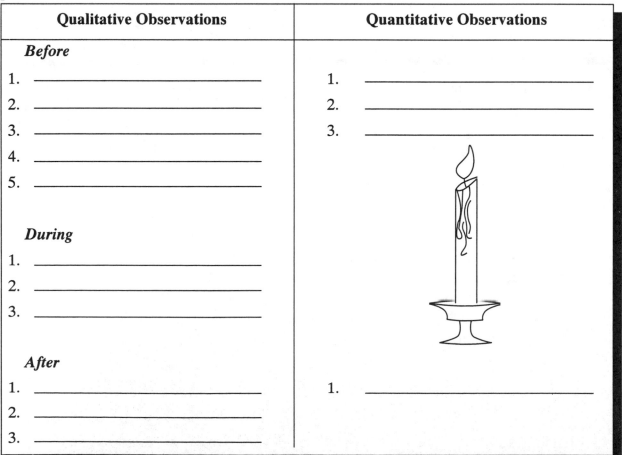

Compare your answers with someone else's or check your answers with the ones that follow. Then proceed to the Self-Assessment.

SELF-CHECK

Below is a list of observations that could be made about a candle before it is lit, during burning, and after it has burned. Some of your observations may be different depending on the particular candle, whether it was new or previously used, and the availability of a sensitive scale. This, of course, is not a complete list. Please see your instructor if you have any questions about the observations you have made.

Qualitative

Before

1. Color: white
2. Slight odor
3. Undetectable taste
4. Cylindrical shape
5. One end flat, other end cone shaped
6. From cone extends a tuft of white, fuzzy, fibrous, soft material composed of strands
7. Each strand is cylindrical and irregularly coiled

During

1. Fibrous strands turn black
2. Flame is elliptical in shape
3. Flame flickers in slight wind
4. Upper part of flame is bright yellow, lower part of flame is dull yellow with a blue margin
5. A puddle of liquid forms in place of the cone
6. Liquid material drips down side of candle; some solidifies on a cooler part of the candle, some drips to the table top

After

1. Color: white
2. Solid irregular in shape
3. Small portion of fibrous strands protrudes from wax
4. Exposed part of fibrous strands is black

Quantitative

1. Mass: 2 g
2. 5 cm long
3. 5 mm diameter
4. Each strand is 0.5 mm in diameter
5. Coil of strands is 1 mm in diameter
6. Coil extends 5 mm above tip of cone

1. Mass: 1 g
2. Height of wax at highest point: 3mm
3. Distance across wax at the widest point:1.5 cm

⇨ *Go* to the observation supply area and pick up a spoon and the effervescent material provided by your instructor.[1] Also obtain a clear container half-filled with water. In this assessment you will be observing the changes that take place as the material dissolves in water.

In the chart below list at least eight observations about the material before placing it in water. At least three of your observations should be quantitative.

Drop the material into the container of water and list at least three qualitative observations about the change while it is occurring.

When the change seems to be complete, record what you observe. At least two of your observations should be quantitative. You may wish to use the spoon to help in your investigation.

Qualitative Observations	Quantitative Observations
Before	
1. _____	1. _____
2. _____	2. _____
3. _____	3. _____
4. _____	
5. _____	
During	
1. _____	
2. _____	
3. _____	
After	
1. _____	1. _____
2. _____	2. _____
3. _____	

Compare your answers with someone else's. Be sure that your observations cover the properties listed in the Self-Assessment answers.

1. Alka-Seltzer tablet or generic equivalent, Bromo Seltzer granules, or "fizzing candy"

SELF-ASSESSMENT ANSWERS

Your observations about the effervescent material before, during, and after the change should have included statements about the following;

Qualitative	**Quantitative**
Shape	Number of particles
Color	Mass
Texture	Size of particles
Smell	
Taste	

IDEAS FOR YOUR CLASSROOM

1. Objects that can be interesting to observe are flowers, fruits, a pine cone, different kinds of leaves, feathers, and dried foods such as cereals.
2. Events such as popcorn popping, making ice cream, making butter or cookies can be delicious as well as informative.
3. A simple drop of water can be fascinating and lead to many challenging questions. Place a single drop of water on a paper towel or ordinary paper. What happens? (Water is attracted by paper fibers and is absorbed.) Place a drop of water on waxed paper. What happens? (The water drop "balls up"—cohesion.) Tip the waxed paper so that the drop moves. Does it roll or slide? How can you find out? (Hint: Sprinkle it with pepper or chalk dust.) Place a drop of water on plastic wrap. Place the plastic wrap so you can look at some printed material by peering through the water drop. What happens? (It magnifies.) Experiment with larger and smaller water drops to see which makes better magnifiers.
4. Observation helps us learn that important changes are taking place.
    a. Seal one nail in plastic sandwich bag and another nail in a plastic sandwich bag with a dampened paper towel. Observe for several days. What differences were observed? Why? An interesting spinoff of this lesson would be to come up with different ways to prevent the nail from rusting.
    b. A similar lesson could be conducted using bread. Students would learn that moisture is important for rotting to take place.
    c. Observing changes taking place with a banana peel in a sealed sandwich bag could lead to interest in how different kinds of foods are preserved. It could lead to an interesting history lesson. How did pioneers preserve food? What did those foods taste like?
5. "Our Senses Depend on Each Other" is an enjoyable lesson.
   Have the students close their eyes so they aren't peeking and hold their noses so they can't smell. Give them a small sliver of apple, then raw potato, then raw onion. Can the students taste the difference? Now let them do it with their noses opened. Now can they tell the difference? Why does the sense of smell help the sense of taste?

6. "Autumn and the Five Senses" would make an excellent theme for a bulletin board and interest center. Fruits and vegetables, the changing color of leaves, and other changes could be observed.

7. "Safety and Our Senses" is another topic that is worthy of teaching. One way to approach this topic is to use sense deprivation. For example, tape could be placed on the ends of the fingers and students could compare how rough, coarse sandpaper feels between taped and untaped fingers. This lesson could be extended by imagining no sense of touch. How would we be protected from sharp objects, heat, sharp blows, or blisters on our feet? Smell warns us when we are breathing something that could be harmful to us. You could use vinegar to simulate noxious gases. Smelling smoke can also warn children of the danger of fire and being burned. The sense of taste can be discussed as something that can warn of danger, but stress the hazards of tasting unknown substances. This would be a good time to discuss with students some of the poisons and poisonous plants found in the home. Sight and hearing are more obvious as signals of danger (e.g. horns, traffic lights, and so on) but students still need to learn about safety. Again, the lesson could be started by having students imagine how they would cope with danger if they had no sense of sight or hearing. The school nurse or a local doctor could be invited to talk to students about eye and ear care and perform sight and hearing tests.

THOUGHT STARTER QUESTIONS

- What do you observe about that object (or event)?
- List as many properties as you can about that object.
- What are all the things you observe directly about . . . ?
- Describe how this object looks, feels, smells, and sounds.

ASSESSING FOR SUCCESS: OPEN-ENDED QUESTION OR SITUATION

Directions to the Students

A. Observe the aquarium[1].

B. Write three observations about the aquarium picture.

1. _____

2. _____

3. _____

Scoring Procedure

1 point for each correct observation.

Acceptable responses include:

- some fish are bigger than others.
- there are two snails in the aquarium.
- the aquarium is not completely full.
- snails are on the bottom of the aquarium.
- the leaves of one plant look like feathers.

Unacceptable responses that are not observations include:

- snails and fish don't like each other.
- more fish could live in the aquarium.
- the small fish are babies of the big fish.
- fish near the top are breathing air.
- the plants will not have enough air.

---

[1] Although an actual aquarium is preferred, a picture of an aquarium may be used. A terrarium or other interesting objects could be substituted as well.

Communicating

To do the communicating activities that follow you will need:

✓ a set of "Sensory Materials" (consisting of a variety of objects to smell, taste, feel, and see)
✓ a set of tangram pieces (A tangram is a Chinese puzzle made by cutting a square into five triangles, a square, and a rhomboid. The pieces are used to form different figures and designs. You will use the set of tangrams on page 269 for activity 3.)
✓ a magnetic compass (optional)

WHY IS COMMUNICATING IMPORTANT?

Our ability to communicate with others is basic to everything we do. Graphs, charts, maps, symbols, diagrams, mathematical equations, and visual demonstrations, as well as the written or spoken word, are all methods of communication used frequently in science. Effective communication is clear, precise, and unambiguous and uses skills that need to be developed and practiced. As teachers we attempt to influence behavior through the written or spoken word. We all have a need to express our ideas, feelings, and needs to others and we begin to learn early in life that communication is basic to problem solving.

GOALS

These exercises will help you learn to communicate ideas, directions, and descriptions effectively and give you practice in using and constructing various methods of communication.

B. C. by permission of Johnny Hart and Field Enterprises, Inc.

PERFORMANCE OBJECTIVES

After completing this set of materials, you should be able to:

1. Describe an object or event in sufficient detail so that another person can identify it.
2. Construct a map showing relative distances, positions, and sizes of objects with sufficient accuracy so that another person can locate a particular place or object using the map.

Activity 1

USING GOOD DESCRIPTORS

Do the following activity alone. Once you have completed this activity by yourself, talk with other students about what they observed about the objects and add to your list of possible descriptors.

⇨ *Go* the supply area and obtain a set of *Sensory Materials*. In this activity you will explore a wide variety of objects displaying several different properties. As you observe these objects think about how you would describe the properties to someone else. Your task in this activity is to generate a list of descriptive words (descriptors) that can be used to effectively communicate what you observe (smell, feel, taste, hear, and see) to others.

Keep in mind you are not attempting to name the objects or describe how you feel about the properties, you are just describing properties.

Smell — Some words to describe how things *smell*

Touch — Some words to describe how things *feel*

Taste — Some words to describe how things *taste*

Sound — Some words to describe how things *sound*

Sight — Some words to describe how things *look*

Just a few possible descriptors are listed in the *Self-check* that follows.

SELF-CHECK ✓

Some words to describe how things —

Smell   sweet, rotten, smoky, fresh, spicy, pungent, strong, moderate, weak, lemony, oily, minty, moldy, woody, sweaty, perfume-like.

Taste   sweet, sour, bitter, strong, moderate, weak, rich, spicy, syrupy, acidic.

Feel   rough, smooth, sandpaper, feathery, slick, cold, hot, warm, cool, rubbery, prickly, sharp, soft, hard, gritty, fuzzy, furry, scaly, cottony, bumpy, oily, waxy, sticky, wet, dry, moist, slippery, leathery, powdery, crumbly, creamy, glassy, jagged, slimy, vibrating.

Sound   loud, moderate, soft, brassy, high, low, medium pitch, sharp, dull, rattle, ringing, muffled, clear, distinct, squeaky, bark-like, scraping, tearing, banging, crashing, dripping, clicking, crinkling, abrupt, continuous, sudden.

Look   colors, shapes, designs, shiny, dull, clear, cloudy, sparkles, bubbly, bright, intense, continuous, interrupted, muted.

COMMUNICATING DESCRIPTIONS

When you describe an object to someone, your purpose will be better served if your communication is an effective one. You can communicate effectively if you:

1. Describe only what you observe (see, smell, hear, and taste) rather than what you infer about the object or event.
2. Make your description brief by using precise language.
3. Communicate information accurately using as many qualitative observations as the situation may call for.
4. Consider the point of view and past experience of the person with whom you are communicating.
5. Provide a means for getting *feedback* from the person with whom you are communicating in order to determine the effectiveness of your communication.
6. Construct an alternative description if necessary.

In the next activity you will communicate a description to someone else. You will then receive feedback on the effectiveness of your communication.

A ctivity 2

DESCRIBING OBJECTS TO OTHERS

Look at the figure that follows:

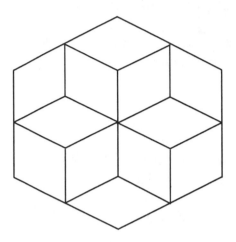

Think about how you might describe this figure to someone in sufficient detail so that he or she could draw it from your description. The artist will need to know what kind of lines to draw, where to place them, and how long they should be. Locate a person to be the artist and do the following:

- Look at the figure again and keep looking at it until you perceive it in a way that is different from how you first perceived it. (There are at least eight different ways to perceive this figure.) The way you describe something to someone else depends on how you perceive it.

- Carefully consider how you will describe the figure to the artist before you begin speaking.

- Without showing the figure to the artist, *effectively communicate* to that person how to make the lines so that their completed drawing looks as much like the original figure as possible.

- The similarity between the figures is a measure of the effectiveness of your communication.

PRACTICING USING COMMUNICATION TOOLS

In order to develop good communication skills, everyone needs ample opportunity to *practice* communicating effectively with others. Children who do little talking in school, who spend hours in front of a TV at home, and who have little contact time with adults have few opportunities to develop good communication skills. While *talking* about what they are learning, children discover new ways to construct their thinking. They also may learn to resolve discrepancies with others in acceptable ways. Talking while they are doing science activities, making entries in journals, recording and organizing data, comparing results, and sharing findings are all activities that help children develop effective ways to communicate.

Learning to use the *tools* of communicating helps children to be able to make good decisions about *how* to communicate observations and ideas. Here are some tools you can use to share what you know:

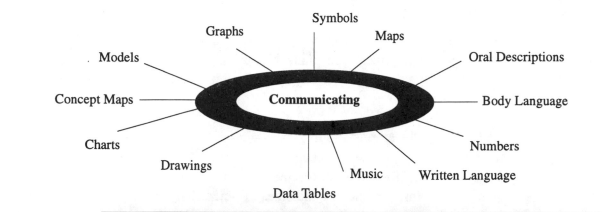

B. C. by permission of Johnny Hart and Field Enterprises, Inc.

GIVING AND FOLLOWING DIRECTIONS

↪ *Go* to the communication supply area and pick up a set of tangrams or cut them out of page 269. Select a partner and be sure that both of you have identical sets of tangrams. Sit with your partner and erect a screen or barrier so you cannot see each other's tangrams. Make a design with your tangrams. (It is not necessary to use all the pieces.) Giving precise directions, tell your partner where to place each tangram so that his or her design looks exactly like yours. How effectively you communicate with your partner will be measured by the closeness of similarity between the designs when your directions are completed. Do this activity a second time and see if your skills of communication improve.

You may wish to identify the specific areas of giving directions which gave you some trouble so that you can practice and improve in these areas. Use the space below to make notes for yourself.

• Describing only what is observed:

• Making descriptions brief:

• Using precise language:

• Communicating information accurately:

• Considering another's point of view or position:

Activity 4

COMMUNICATING WITH MAPS

Some directions can be effectively communicated only with the use of maps. (Actually mathematical formulas, patterns, guides, floorplans, blueprints, photographs, schematic drawings, and descriptions are all maps.) A map is any symbolic representation. To be useful a map must have:

1. a title, telling what the map is about.
2. symbols, representing places or objects.
3. a key, telling what each symbol represents.
4. a scale, showing relative distances and sizes of objects.

Suzanne wanted to tell her classmates about her trip to one of the islands of Hawaii. She drew a map of the island and labeled all the places on the map that she visited. Then she used the map to show the class where she had been and what she did there. By placing numbers along the top of the map and letters along the side of the map, she found it was easier to tell her classmates where places were located. Diamond Head (a once active volcano), for example, is located at about H-7. That is the place on the map close to where lines drawn from H and 7 would cross. Now it is your turn.

Following Suzanne's map is a list of fun things she did and places she went on Oahu. For each place listed, locate its position on the map and identify its location by naming the appropriate letter and number on the map.

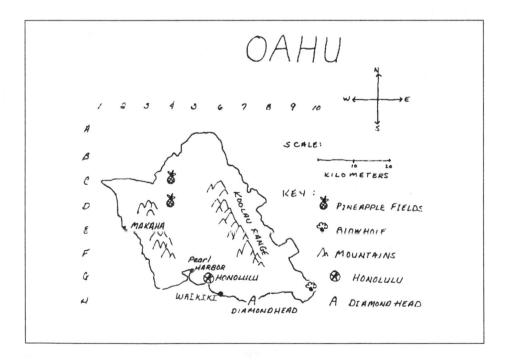

1. _____ visited the pineapple fields.

2. _____ saw the monument at Pearl Harbor

3. _____ saw the geyser of water erupting from the Blow Hole

4. _____ went surfing at the beach of Makaha

5. _____ went swimming and snorkeling at Waikiki Beach

SELF-CHECK ✓

1. C—4 and D—4
2. G—5
3. G—10
4. E—2
5. H—6

Self -Assessment Communicating ✍

1. Place three objects somewhere in the room and write a description of one of the objects.
2. Construct a map of the room so that a person can find each of the objects (you may wish to use a magnetic compass.)

SELF-ASSESSMENT ANSWERS

Give your map and description to someone in your class and have them locate and recognize the object desired. If the person using your map and description finds and recognizes the object in just one attempt then you know your communication is an effective one. Give yourself a pat on the back!

IDEAS FOR YOUR CLASSROOM

1. Although it was not covered in this chapter, writing is a very important communication skill. Good clear writing like clear verbal communication must be practiced. One way this can be practiced in science is to have students write what they are learning, especially during activities. You may want to adopt a format similar to the one below:

    a. We were studying _____

    b. We did _____

    c. We observed _____

    d. We learned _____

2. A game that will help students' descriptive skills is similar to one earlier in the chapter. Place two or three students in each group. Let one student pick out an object in the room and describe it to the others. (The description should be observations rather than function, that is, it is red rather than you write with it.) When the object is correctly identified, then another student gets a turn.

3. An interesting application of communication skills to social studies would be to obtain an old map of your geographic area and a current map. If you are lucky enough to get a map of your own area when your state adopted its seal (excepting Hawaii and Alaska), you get a bonus. Compare the maps. What changes have taken place? What used to be where you are now? Where is your home? Now for the bonus. . . Your state seal is a piece of communication. When it was adopted, the people of the state were trying to communicate the state's important qualities at that time. Have these qualities changed? If you were going to design a state seal, what would you include now?

4. Using our Resources

    a. *Where does it originate?* is an interesting question for a study. Bulletin boards and activities could be used to help students gain an appreciation and understanding of how we use our environment to meet our needs. Many children need to learn that stores are not the sources of eggs, milk, pencils, baseball bats, and clothes. In a study of the sources of foods, chains could be constructed to show events between the producers and consumers. (The chains could vary in level to fit the sophistication of the students.) If desired, steps could be added to illustrate processing, transportation, sales or any other event between the raw product and the consumer.

      Where does electricity originate? When you turn on a switch, does the electricity come from a local plant or one far away? What kind of plant generates the electricity? What kinds of energy transformations take place between its production and use? As you use electricity, what kinds of energy transformations take place?

    b. *Where does it go?* is another question worth studying. What happens to things we "throw away"? This could lead to a study of waste disposal and some of its problems as well as to a study of recycling. While *What happens when I flush the toilet?* may not be one of the burning questions in your life, it might be worth exploring.

5. *Gossip* is an interesting game that helps improve communication skills. You could start the game by having the students form a circle. Then you give a short written message to one of the students. The student reads the message and whispers it once to his or her neighbor, who in turn passes the message along verbally. When the message has gone around the circle, the last student says the message aloud. Compare the original message with it. This could lead to a discussion of how we receive information. Perhaps it would be possible to visit a radio or television station or newspaper office. If that's not possible, maybe they have a speaker that could visit your class.

6. *Kidnap* is another game that can improve observation and communication skills. Have three students, one victim and two villains, perform the following skit. Dressed in special clothes, the masked villains kidnap the victim (also dressed in unique garb) quickly from the room. Have the class write a brief eyewitness account, describing the event and the descriptions of the victim and kidnappers. Compare the results. You might have a person from the police visit the class to talk about accuracy of communication and being an eyewitness

7. Labels communicate! Have your students become label readers. Find out the contents of junk foods or any other food that comes in a container. What other products have labels? What does the label tell about the product?

8. Advertisements are another form of communication. Have your students study different advertisements from different sources: television, magazines, newspapers, etc. What are they communicating? Are there hidden messages?

THOUGHT STARTER QUESTIONS   ?

- What words would you use to describe this object so that someone else can identify it?
- Describe everything you observed as completely as you can.
- Using good descriptive words tell everything you observed about this object (or event.)
- What method would you use to communicate to someone else what you observed?

ASSESSING FOR SUCCESS: PENCIL & PAPER TEACHER OBSERVATION CHECKLIST

Observational checklists may be used to assess students acquisition and use of specific skills. Use a check mark or date notation to document appropriate process skill behaviors as you observe them. Compare this paper and pencil approach to assessment with a new technological method for recording teacher observations that is illustrated in Chapter 11.

Student Name	Uses appropriate vocabulary to describe objects and events	Communi-cates clearly	Listens to others	Selects and uses appropriate communica-tion tools

Classifying

Sight

Smell

Sound

Taste

Touch

Most of the items you will need for classification are included in the activities themselves. For some activities you will need the following:

- ✓ a set of 6 assorted buttons (numbered 1, 2, 3, 4, 5, 6)
- ✓ an assortment of information panels from cereal boxes
- ✓ 6 peanuts in the shell
- ✓ an assortment of pasta shapes

WHY IS CLASSIFICATION IMPORTANT?

For us to comprehend the overwhelming number of objects, events, and living things in the world around us, it is necessary to impose some kind of order. We impose order by observing similarities, differences, and interrelationships and by grouping objects accordingly to suit some purpose. The basic requirement of any system of grouping is that it must be useful. Think of the number of classification schemes of which you are a member. Scientists classify you (human) for the purpose of study; the telephone company classifies you so that you can receive phone calls; your employer classifies you according to the work you do. Think of the ways the government classifies you (by sex, age, income, and so on.) There are many classification systems you use almost daily: the *yellow pages*, the classified section of the newspaper, the Dewey Decimal System in libraries, the systems for arranging items in grocery and department stores, and many more. As a teacher, you may rank students according to how much they know. Further, it is important to remember that classification is the process skill central to concept formation.

B. C. by permission of Johnny Hart and Field Enterprises, Inc.

GOAL

The purpose of these exercises is to help you learn to classify objects and events on the basis of observable characteristics.

PERFORMANCE OBJECTIVES

After completing this set of activities, you should be able to:

1. Given a set of objects, list observable properties that could be used to classify the objects and construct a binary classification system for each property.
2. Given a set of objects, construct a multistage classification system and identify the properties on which the classification is based.
3. Given a set of objects, identify properties by which the set of objects could be serially ordered and construct a serial order for each property.

Activity **1**

CONSTRUCTING A BINARY CLASSIFICATION SYSTEM BASED ON OBSERVABLE PROPERTIES

In a binary classification system the set of objects is divided into two subsets on the basis of whether each object has or does not have a particular property. To construct a binary classification system you must first identify a common property that only some of the objects have. Then group all the objects displaying that property in one set and all the objects not displaying that property in another set. For example, biologists classify living things into two groups: animals and plants (plants being the group *not* displaying animal properties). Scientists further classify animals into two groups: those with backbones and those without backbones. When constructing a binary classification system be certain that all the objects in the original set will fit into one and only one of the two subsets. This is shown in the following activities.

Look at the set of *creatures* and observe their similarities and differences. In the left hand column of the chart provided, list at least three observable properties by which the creatures can be grouped into two subsets. In the *Yes* column write the numbers of the creatures that have the property that you have identified. In the *No* column, write the number or numbers of the creatures that lack the property that you have identified. Examine the following example. Binary classification is essentially a yes/no grouping system. Yes, they have the properties; no, they don't have the properties.

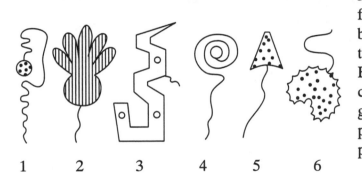

1   2   3   4   5   6

This drawing was reproduced from the Elementary Science Study Unit, ATTRIBUTE GAMES AND PROBLEMS. Copyright © 1984 by Delta Education, Hudson, NH.

Observable Properties	**Yes**	**No**
1. Speckled body (example)	1, 5, 6	2, 3, 4
2.		
3.		
4.		

Notice, as in the example, that for each property the subsets accommodate all the objects in the original set and that every object can be assigned to one of the subsets. Check to be certain your work in the preceding chart meets these two important requirements.

Compare your answers with someone else's or check your observations with those that follow.

SELF-CHECK ✓

Some of the properties you may have identified are as follows. Be certain that for each property the subsets accommodate all the objects in the original set, and that every object can be assigned to one and only one subset.

Observable Properties	Yes	No
1. (example) speckled body	1, 5, 6	2, 3, 4
2. Round body	1, 4	2, 3, 5, 6
3. Body with scalloped margin	6	1, 2, 3, 4, 5
4. Striped body	2	1, 3, 4, 5, 6
5. Body with circles	3	1, 2, 4, 5, 6
6. 3/4 of tail coiled	4	1, 2, 3, 5, 6

MAPPING AN ORGANIZATION SCHEME

A chart such as the one shown in Activity 1 is a useful communication tool. Notice that this chart is binary in nature because it sorts items into either one category or another according to whether or not it possesses a certain characteristic.

Another powerful communicator is a graphic organizer or *map*. Notice that the map shown here is also binary because the original set is split into only two subsets. The map clearly identifies which properties are being used to sort the items and indicates which items belong in each subset.

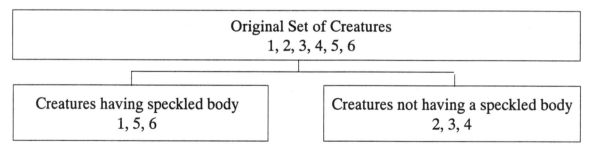

You'll be asked to draw a similar classification map in the next activity.

 *A*ctivity *2*

USING SEVERAL PROPERTIES AT ONCE

Binary classification also can be used when you wish to group objects that have more than one property in common.

⇨ *Go* to the supply area and pick up a set of buttons.

Those buttons having several properties in common may be sorted into one group and those buttons not having those properties may be sorted into the other group. The first group, for example, may contain all the buttons that are *white and have four holes*. The second group will contain all the buttons from the original set that do not possess *both* of these properties.

1. Using your set of buttons, what are some ways the buttons can be grouped using two or more characteristics?

   • _____

   • _____

   • _____

2. Construct a map to show a binary classification scheme using several properties of the buttons. The box showing the original set has been drawn for you but you'll need to write in the button numbers. Then complete the scheme making sure it is binary and that each subset contains the names of the properties and the numbers of the buttons possessing those properties.

Original Set of Buttons

See the Self-check on the next page.

SELF-CHECK ✓

Answers will vary. They could include:

1. red buttons with two holes
2. white buttons with four holes
3. blue buttons with rough edges
4. the answers could be more subtle. For example, one set could be red and blue buttons with two holes and smooth edges.

Your map will differ according to the type of buttons you used and the properties you chose, but here is an example.

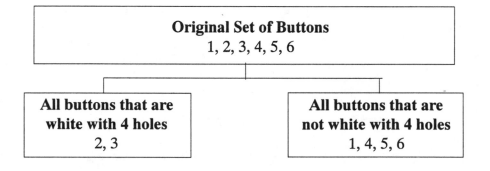

Another way to check your map is to give it and your buttons to someone else. If they are able to use the map to replicate your several property classification scheme, it must be a good one.

A good way to sharpen your binary classification skill is through the *Button Game*. Get a partner. The rules for this game are quite simple.

1. Make up a rule for grouping the buttons into two sets.
2. Group the buttons according to your rule.
3. See if your partner can guess the rule.
4. When your partner guesses this rule, he or she gets to make up the rule, group the buttons, and then let you guess.

Activity 3

CONSTRUCTING A MULTI-STAGE CLASSIFICATION SYSTEM

Performing a binary classification on a set of objects and then again on each of the subsets results in a classification system consisting of layers or stages. When objects are classified in a binary way again and again, a hierarchy of sets and subsets is created, called *multi-stage classification*. As in a binary scheme, subsets are determined by sorting objects that have a particular property from those that do not have that property. Animals, for example, are classified as either having backbones or not having backbones. Those having backbones can be further classified as either having hair or not having hair.

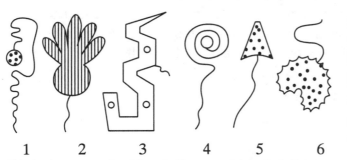

1   2   3   4   5   6

This drawing was reproduced from the Elementary Science Study Unit ATTRIBUTE GAMES AND PROBLEMS. Copyright © 1984 by Delta Education, Hudson, NH.

Because you are already familiar with the *creatures* in Activity 1, we will use them to illustrate how a multi-stage classification system is constructed. Notice that the scheme identifies the observable properties by which the items are sorted and that each item associated with the properties is identified.

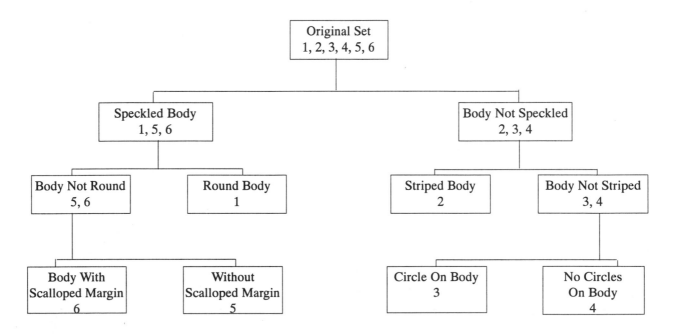

In addition to the characteristics already discussed, a multi-stage classification system has the following features:

1. Several other schemes may be possible depending upon which observable properties are used for grouping.
2. When each object in the original set is separated into a category by itself, the scheme is complete.
3. A unique description of each object can be obtained by listing all the properties that the object has. In the above scheme, for example, creature 6 can be distinguished from the other creatures in the original set by listing its properties; body is speckled, not round, but has a scalloped margin.

Now it's your turn to construct a multi-stage classification scheme. Use the same numbered button assortment as in Activity 2. Complete the system started below. In each box, indicate the property such as color, number of holes, size, and so on that you used to make the grouping. Be sure to carry the scheme through to completion. You may need to add more boxes. When the scheme is completed, each button will be in a box by itself.

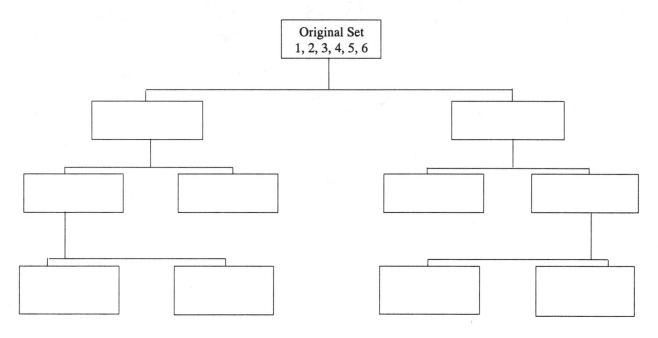

SELF-CHECK

There are many possible schemes depending on the buttons' characteristics and the order in which you selected them. If in doubt, ask your instructor.

For additional practice, do this classification activity. You'll need sharp observation skills!

A ctivity 4

MULTI-STAGE CLASSIFICATION ADDITIONAL PRACTICE

⇨ *Go* the supply area and obtain a handful of peanuts (about 6) in the shell.
Observe your set of peanuts carefully.
Construct a multi-stage classification scheme for your peanuts making sure your scheme fulfills these requirements:

- The items in the original set are identified.
- Each sorting action is binary.
- Each subset is identified by one or more properties and the items possessing those properties.
- The scheme is continued until each item is in its own subset.

You should number your peanuts . . .

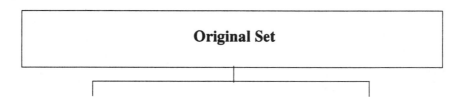

Original Set

SELF-CHECK ✓

The best way to check your map is to give it and your set of peanuts to someone else and have them use it to classify the peanuts.

A ctivity 5

SERIAL ORDERING

It sometimes is necessary to order objects according to the extent to which they display a particular property. Depending upon the purpose of the classification, objects may be ordered on the basis of size, shape, color, or a variety of other characteristics. In a hardware store, nails are ordered on the basis of size. Paints can be arranged according to size of can or color. Clothing stores use size to arrange merchandise in serial order.

⇨ *Go* to the supply area and obtain the information panels from cereal boxes. Examine them and identify three properties by which these panels could be arranged in order.

1. _____

2. _____

3. _____

SELF-CHECK ✓

Answers will vary. They could include:

1. number of calories per serving.
2. amount of iron per serving.
3. amount of thiamin per serving.
4. amount of protein per serving.

One way to graphically show a serial ordering is with an arrow that is labeled to indicate an increase or decrease in the extent to which items display a certain property. You might, for example, serial order students in your class according to height. That serial ordering might look like this:

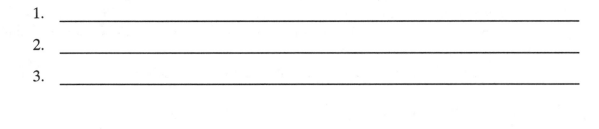

Serial order the panels of cereal boxes according to one of the properties you identified earlier. Represent the serial order on the arrow shown. Be sure to label the *property* that all the items display, the *degree* to which they display the property, and each item's *position of rank* along the continuum.

_____ ──────────────────────▶ _____

_____

SELF-CHECK

Your serial ordering will be different depending upon the cereals you used and the property you selected but here is an example.

Least →──────────────────────────→ **Most**

Shredded Wheat	Rice Krispies	Wheaties	Grape Nuts
4%	10%	25%	45%

Amount of Iron per Serving (% RDA)

Self-Assessment Classifying

For this test you will construct: (a) a binary classification system; (b) a multistage classification system; and (c) a serial order classification system for a set of pasta shapes.

↪ **Go** to the supply area and obtain a set of pasta shapes (shell, spiral, elbow, wheel, tube, bowtie.) **Your instructor may be using other pasta shapes.**

a. In the chart below identify at least three observable properties by which these shapes could be classified in a binary classification system. In the proper column indicate which shapes have or do not have each property.

Observable Properties	Yes	No
1.		
2.		
3.		

b. On a separate piece of paper construct at least one multi-stage classification system for the pasta shapes and carry it through to completion. In each box identify the property used for grouping and list the name of each shape with that property.

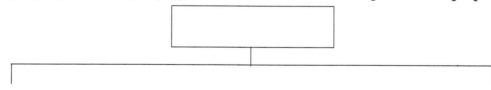

c. Identify a property by which the pasta pieces can be serial ordered. Then serial order the pieces on the basis of that property. In the space below draw and label an arrow that accurately communicates how you serial ordered your pasta pieces.

SELF-ASSESSMENT ANSWERS

a. Some observable properties on which binary classification systems for these pieces of pasta could be based are listed below (our pasta may differ slightly from your pasta):

Observable Properties	Yes	No
cylindrical shape	tube, wheel, elbow	spiral, shell, bowtie
twists or turns	tube, elbow, spiral	shell, wheel, bowtie
has compartments	wheel	spiral, shell, elbow, bowtic, tube
ribbed surfaces	tube, wheel, shell	spiral, elbow, bowtie

b. One possible multi-stage classification scheme for a set of pasta pieces is shown below. Several other schemes are possible depending on which properties are used for grouping. Keep in mind your pasta may differ from ours. If you have questions about your classification scheme, compare it with someone else's or see your instructor.

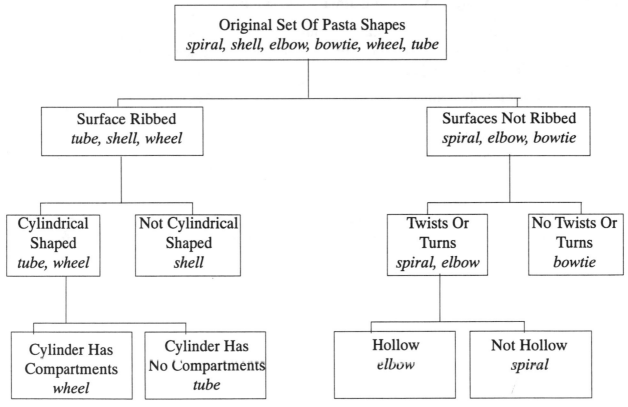

c. A couple of ways of serial ordering your pasta pieces are shown below (again our pasta may differ from yours):

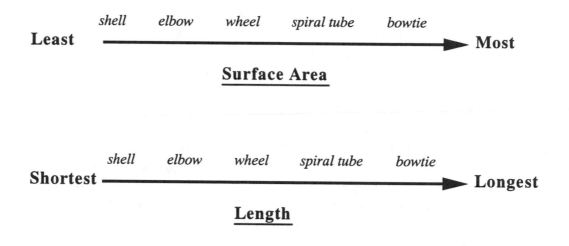

	shell	elbow	wheel	spiral tube	bowtie	
Least						→ **Most**

Surface Area

	shell	elbow	wheel	spiral tube	bowtie	
Shortest						→ **Longest**

Length

IDEAS FOR YOUR CLASSROOM

1. Water play is an enjoyable activity for young learners. In one set of activities, students can compare objects that sink with objects that float. (This is a building block for later concepts of density and buoyancy.)
   In another set of activities, students can predict the serial order of the volumes of empty containers and check their prediction by filling the containers, one with the others. (In doing this, they learn how to relate sizes and shapes with volumes of containers.)

2. Liquids vary in density and this affects their buoyant properties. A neat puzzle for older children is to present them with two containers of water: one in which a hard boiled egg is floating and the other in which the hard boiled egg is resting on the bottom. (The only difference between the glasses of water is that one contains a few tablespoons of dissolved salt.)

3. Animals are grouped on the basis of similarities and differences. Ways in which they may be grouped include: how they obtain their food; whether they lay eggs or bear live young; by their habitat (parks, forests, oceans, ponds, and so on); or how they protect themselves. Have you ever wondered why a zebra had stripes?
   Collect pictures of various animals and explore the various ways they can be classified using binary classifications. If your class feels really ambitious, they may construct a multistage classification system for animals. (You may want to limit their efforts to local kinds of animals because the entire animal kingdom is a gigantic multistage classification system.)

4. Plants can be used for binary, multistage classification and serial ordering. Plant leaves are excellent for binary and multistage classification. Trees, for example, could be serial ordered by the distance around, height, age (by counting the rings) and a number of other ways.

   One interesting project would be to investigate how plants disperse their seeds: some rely on the wind; some need animals to eat their fruit; some "hitch" rides on animals; just to mention a few. Further, examine the similarities and differences of plant seeds dispersed in the same manner. Maple seeds and dandelions are both spread by the wind but their methods are different. Dandelion seeds could be likened to parachutes and maple seeds could be compared to helicopters.

5. Is it natural or man-made? Some fabrics from which clothes are made occur naturally while others are manufactured from other products like petroleum.

6. Nutrition is a natural unit to practice classification skills. Besides studying the food groups, you can have your students read the sugar content of various cereals and measure out equivalent amounts of sugar. (Sand or salt may be substituted to give the same visual effect at a lower cost.) Once the amounts have been measured out, they can be ordered.

THOUGHT-STARTER QUESTIONS

- How is this object like that one?
- How are these objects alike; how are they different?
- Sort these items according to their properties.
- Put these objects in order and explain the order
- Group these objects by color (size, texture, smell, taste, or sound they make.)
- Classify this set of objects in as many different ways as you can.
- How would you construct a multi-stage classification scheme for this collection of objects?

ASSESSING FOR SUCCESS: PERFORMANCE TASK

Preparation

The amount of materials needed will depend on how many students are doing the task at the same time. The most economical strategy is to use stations through which students can rotate. Re-sealable clear plastic bags, paper or plastic plates and a variety of beans are needed. When selecting beans, choose beans of different shapes, colors, and sizes.

Directions to the Student

1. Check your materials.
   - Bag of beans
   - Paper or plastic plate
2. Empty the bag of materials onto the plate.
3. Observe the beans. Put the beans into two groups so that something is the same about all the beans in each group.

   * Write one way all the beans you put into group 1 are the same.

   _____

   * Write one way all the beans you put into group 2 are the same.

   _____

4. Put the beans all back together on the plate. Observe them again. Think of another way to put the beans into two groups so that there is something the same about all the beans in each group. Put the beans into two groups.

   * Write another way all the beans you put in group 1 are the same.

   _____

   * Write another way all the beans you put in group 2 are the same.

   _____

5. When you are done, put the beans back into the bag.

Scoring Procedure

For #3   Student identifies a property that all beans in group 1 have in common. Sample acceptable answers are: small, large, long, round, dark, brown, not green, broken, whole, and so on. For group 2, students use any property that they did not use for group 1.

For #4   On both items, students use a property (see sample answers above) not previously used.

Measuring Metrically

CHAPTER 4

To do the measuring activities that follow you will need:

- ✓ a meter stick
- ✓ a metric ruler
- ✓ an equal-arm balance
- ✓ a set of masses
- ✓ 20 centicubes
- ✓ a baby food jar (about 140 mL)
- ✓ 3 large sinkers (about 28 g each)
- ✓ a liter container
- ✓ 4 containers in various sizes and shapes
- ✓ a graduated cylinder
- ✓ 5 marbles
- ✓ 4 washers - about this size
- ✓ ice cubes
- ✓ a Celsius thermometer

For the self-assessment you will need a *Measurement Test Packet* containing:

- ✓ carpenters' nail
- ✓ 10 plastic buttons

WHY IS METRIC MEASUREMENT IMPORTANT?

How much? How far? What size? How long? How many? How fast? These are questions with which we deal every day and we need to be able to handle them with ease. Well developed skills in measuring are essential in making quantitative observations, comparing and classifying things around us, and communicating effectively to others. The change to the metric system of measurement should not be viewed as a problem but rather as a solution to many problems. The metric system gives us easy to learn units for everyday use, and multiply-

B. C. by permission of Johnny Hart and Field Enterprises, Inc.

ing and dividing are relatively easy operations since the metric system is in base ten. Our conversion to the metric system will also give us uniformity with other countries with which we trade and communicate.

GOALS

In these exercises you will learn and practice skills needed to do measurements in the metric system. As you develop these skills you should begin to think metric.

PERFORMANCE OBJECTIVES

After completing this set of activities, you should be able to:

1. Select the appropriate metric unit for measuring any property (length, volume, temperature, mass, and weight) of a given object.
2. Given a set of metric units, state equivalent metric measures using prefixes (perform conversions within the metric system).
3. Measure the temperature, length, volume, mass, or force of any object to the nearest 0.1 unit.

A ctivity 1

MEASURING METRICALLY

The official name of the metric system is Systeme Internationale d'Unites (international system of units), usually known simply as *SI*. The term metric comes from the base unit of length in the system, the *meter*. The meter was originally defined as one ten-millionth of the distance from the equator to the north pole along a meridian that passes through France.

Two spellings of meter (and also liter) are recognized in the United States—meter or metre and liter or litre. Internationally, metre and litre are preferred, but in the United States the spellings of these units are most commonly meter and liter. Although most elementary science and mathematics textbooks and activity books use the spelling meter and liter, you should be comfortable with both spellings and acquaint your students with these differences.

Next to each measurement shown below write whether it should be measured in meters, liters, or kilograms.

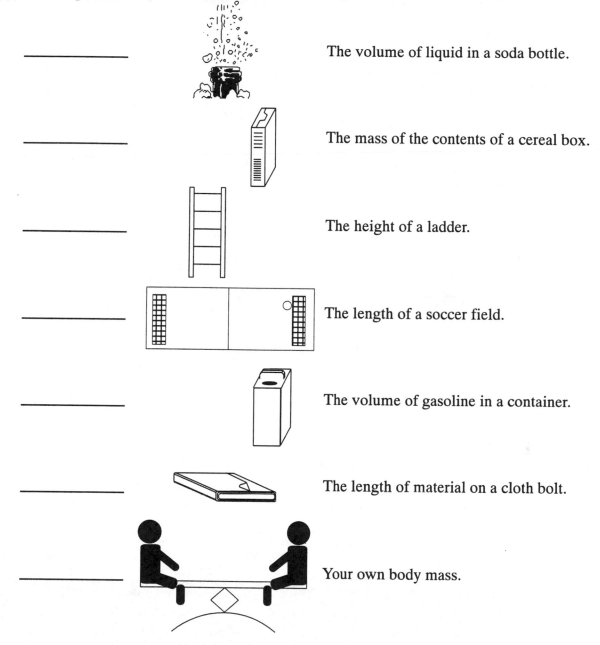

_____ The volume of liquid in a soda bottle.

_____ The mass of the contents of a cereal box.

_____ The height of a ladder.

_____ The length of a soccer field.

_____ The volume of gasoline in a container.

_____ The length of material on a cloth bolt.

_____ Your own body mass.

SELF-CHECK

soda: liter
cereal: kilograms
ladder: meters
soccer field: meters
gasoline: liters
cloth: meters
body: kilograms

Activity 2

METRIC PREFIXES

One day you'll be saying, "give them a centimeter and they'll take a kilometer," and, "a gram of prevention is worth a kilogram of cure," as strange as the language may seem now, one of the advantages of the metric system is its terminology. Instead of remembering conversions like 12 inches equals one foot equals one-third yard equals 1/5280 mile, the metric system uses prefixes to indicate larger and smaller quantities. By adding prefixes to the basic units of measure (meter, liter, and gram), you can indicate larger and smaller quantities.

The most frequently used prefixes are:

Prefix	Prefix Meaning
kilo _____	1000 times the base unit
hecto _____	100 times the base unit
deka (deca) _____	10 times the base unit
deci _____	1/10 of the base unit
centi _____	1/100 of the base unit
milli _____	1/1000 of the base

As you can see. . . kilo
                    hecto   } . . . make the base unit larger
                    deka

                    deci
                    centi   } . . . make the base unit smaller
                    milli

Two other prefixes that are becoming more frequently used are *mega* and *micro*. *Mega* means one million times the base unit and *micro* means one millionth of the base unit. When might you use these prefixes? You may wish to place these prefixes in the chart above.

The metric system is much like our monetary system because both are decimal (based on ten) systems. For example:

Just as . . .	1 dollar	= 10 dimes or 100 cents,
so . . .	1 meter	= 10 decimeters or 100 centimeters

You will be using the prefixes with each of the metric base units so spend a few minutes right now getting to know them and their meanings. Then do the practice exercise that follows. Use your knowledge of the prefixes and their meanings to answer the following questions.

1. One meaning of decimate is to reduce something to _____ its original size.

2. A decapod is an animal which has _____ legs.

3. In the decimal system, 0.3 is read as three _____ .

4. A hectometer equals _____ meters

5. A kilowatt equals _____ watts.

6. If a centipede has as many legs as its name implies, each leg is _____ (what part) of the total number of legs.

7. On the centigrade (Celsius) temperature scale each degree is _____ (what part) of the scale.

8. In a millennium, each year is _____ (what part) of the total period of time.

9. A mill is _____ (what part) of a dollar.

Check your answers with the ones in the Self-check.

SELF-CHECK ✓

1. decimate: 1/10 (deci) the original size
2. decapod: 10 (deka or deca) legs
3. decimal: tenths (deci)
4. hectometer: 100 (hecto) meters
5. kilowatt: 1000 (kilo) watts
6. centipede: 1/100 (centi)
7. centigrade: 1/100)centi)
8. millennium: 1/1000 (milli)
9. mill: 1/1000 (milli) of a dollar

Activity 3

MEASURING METRIC LENGTHS

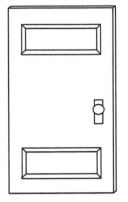

The meter is the basic unit for measuring length in the metric system. A meter is about the distance from the floor to a doorknob.

The symbol *m* is used to for meter.

⇨ *Go* to the supply area and pick up a meter stick. Carefully observe the length of the meter stick, then close your eyes and try to picture that length in your mind. Look around the room and try to select some things which you think are about one meter in length. Then use the meter stick to check your estimations. When you are through you should have a good picture in mind of how long a meter really is.

If you want to learn to *think metric*, always estimate the size of objects before you actually measure them. If your estimates are too large or too small when compared to the actual measurements, you have the opportunity to adjust your thinking about the size of things in metric units. In time, the habit of estimating first and then measuring will give you the ability to *think metric*.

Estimate the following lengths to the nearest meter and record them in the column labeled *estimate*. Then *measure* each of the lengths to the nearest meter and record the measurements in the *measure* column.

	Estimate	Measure
Length of instructor's table		
Width of the instructor's table		
Width of the doorway		
Height of the doorway		
Distance from floor to window sill		

If your estimates and measurements are fairly close, you are beginning to think metric!!
Compare your measurements with someone else's.

You may already have noticed that it may be difficult to measure distances much longer or shorter than one meter using a meter stick. Distances longer or shorter than one meter can be described by using the meter prefixes.

Distances much shorter than a meter can be measured using a metric ruler.

Now examine the meter stick and find the millimeter, centimeter, and decimeter marks on it.

The *millimeter* marks are about as wide as the wire in a paper clip. It is about the distance between two legs of the letter *m*. It takes 10 millimeters to make a centimeter. Millimeter is symbolized *mm*. Although the English system of measurement uses abbreviations with periods (in., yd., oz., lb.), the metric system uses symbols without periods. The only exception occurs when a symbol is used at the end of a sentence.

The *centimeter* marks are about the same length as the width of a paper clip or the width of your little finger. It takes 100 centimeters to make a meter. Centimeter is symbolized *cm*.

The *decimeter* is a little longer than the width of your hand or about the length of a new piece of chalk. It takes 10 decimeters to make a meter. Decimeter is symbolized *dm*.

Use the following scales to compare these lengths. Again, try to picture in your mind how long millimeters, centimeters, and decimeters really are.

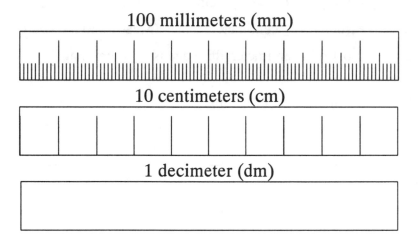

Now let's see if you can put what you have learned to use.
Suppose you are asked to measure this line:

You should carefully lay your metric ruler along the line like this:

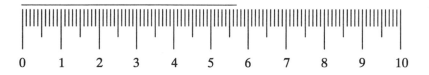

Notice that the line measures more than 5 but less than 6 centimeters. If you were to measure this line to the nearest centimeter, you would say it is 6 cm long since it is closer to 6 cm than 5 cm. More precisely the line measures seven millimeter marks beyond 5 cm. Rather than saying 5 cm, 7 mm you should say 5.7 centimeters, or 57 mm. It is considered bad form to mix metric units. When measuring you will have to decide, or be told, how precisely you should measure.

For practice. . .

On the scale below, identify and label 1 mm, 1 cm, and 1 dm. Then complete the following metric statements.

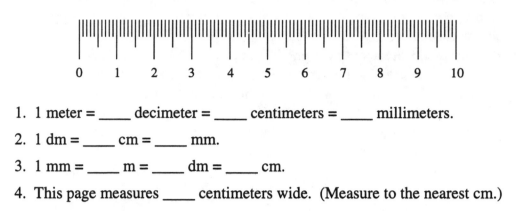

1. 1 meter = _____ decimeter = _____ centimeters = _____ millimeters.

2. 1 dm = _____ cm = _____ mm.

3. 1 mm = _____ m = _____ dm = _____ cm.

4. This page measures _____ centimeters wide. (Measure to the nearest cm.)

5. To the nearest 0.1 of a centimeter, this page is _____ cm long.

Check your answers with the ones following.

SELF-CHECK

1. 1 m = 10 dm = 100 cm = 1000 mm
2. 1 dm = 10 cm = 100 mm
3. 1 mm = 1/1000 m = 1/100 dm = 1/10 cm or = 0.001 m = 0.01 dm = 0.1 cm
4. 20 cm (perforation to edge)
5. 27.5 cm

What about measuring longer distances?

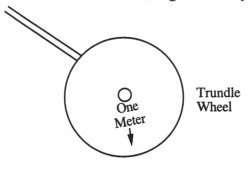

Trundle Wheel

One Meter

Distances longer than a few meters can be measured using a metric tape or wheel calibrated in meters, centimeters, and millimeters. Students have more fun, and learn more, using measuring devices they have made themselves.

Distances of several meters may be measured in dekameters or hectometers, although these units are not frequently used.

The *dekameter* is a measure of length equal to 10 meters. Its symbol is *dam*.

The *hectometer* is a measure of length equal to 100 meters. Its symbol is *hm*.

A race covering a distance of 500 meters is more likely to be called a 500 meter race than a 50 dekameter race or a 5 hectometer race.

For long distances . . .

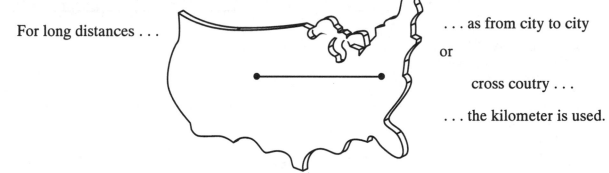

. . . as from city to city

or

cross coutry . . .

. . . the kilometer is used.

The kilometer is a measure of length equal to 1000 meters. Its symbol is km. Kilometer is pronounced **kil** o meter, not ki **lom** eter. In fact, the rule for pronouncing all metric units is to place the primary accent on the first syllable. It would take about 15 minutes to walk a kilometer. The length of nine football fields placed end to end would be about one kilometer. Kilometers can be measured using the distance measuring gauge, called an odometer, in a car. Maps drawn to metric scale can help you determine long distances.

For practice. . .

1. 2 km = _____ m

2. 3 km = _____ m

3. 4000 m = _____ km

4. 21,000 m = _____ km

5. 2 dam = _____ m

6. 500 m = _____ hm

7. 1 km = _____ dam

8. 1 km = _____ hm

Fill in the missing information in the following chart. The middle column asks you to identify the instruments used to measure each distance and the right column asks you to name the units in which each distance is measured.

If you want to measure . . .	You should use a . . .	And measure in . . .
Long distances (as from place to place or city to city)		
Average distances (as short as one step or as long as several football fields)		
Short distances (as short as the width of a finger or as long as a step)		
Very small distances (less than the width of a finger) or *very precise measurements* (accurate to a very small unit)		

Compare your answers for the practice exercise with those on the following chart.

SELF-CHECK

1. 2 km = 2000 m
2. 3 km = 3000 m
3. 4000 m = 4 km
4. 21,000 m = 21 km

5. 2 dam = 20 m
6. 500 m = 5 hm
7. 1 km = 100 dam
8. 1 km = 10 hm

If you want to measure . . .	You should use a . . .	And measure in . . .
Long distances (as from place to place or city to city)	metric odometer (a gauge on a car) or a map (with a metric scale)	kilometers (km)
Average distances (as short as one step or as long as several football fields)	metric stick, metric tape, or metric measuring wheel	meters (m)
Short distances (as short as the width of a finger or as long as a step)	metric ruler	centimeters (cm)
Very small distances (less than the width of a finger) or *very precise measurements* (accurate to a very small unit)	metric ruler	millimeters (mm)

A ctivity 4

MEASURING METRIC MASSES

First a word about mass and weight . . .

There is a very definite difference between mass and weight—while weight refers to how heavy an object is, mass refers to how much stuff, or matter the object is made of. The problem lies in the fact that people often talk about weight when they really mean mass. Consequently, in the everyday implementation of the metric system, mass and weight are being treated as if they were the same; that is, they both use the gram as their base unit. Where it is not necessary for you to make the distinction between mass and weight, the following discussion about mass may also apply to weight. For a more scientific approach to measuring metric weights, see *Measuring Forces*.

The metric base unit of mass (weight) is the gram. How much is a gram?

 . . . The mass of the water a cube this size could hold is 1 gram (g).

. . . Hold a nickel in your hand and try to get a *feel* for its mass. Its mass is about 5 grams.

Masses larger or smaller than the gram can be described using the prefixes with the basic unit. The *kilogram* is the unit most used in measuring large masses. One kilogram is equal to 1000 grams and is the mass of one liter of water. An average man might mass about 80 kg. Very large masses are measured in *metric tons*. One metric ton is the mass of 1000 kilograms.

Masses smaller than the gram are measured in *milligrams*. To get an idea of how small one milligram is, pick up a postage stamp and think about the fact that it masses about 20 milligrams.

To find the mass of objects, we use the equal-arm balance, like the one pictured below. We do so by placing the object of unknown mass on one side and balancing it with objects of known mass on the other side.

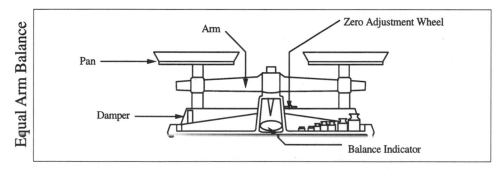

Equal Arm Balance

Arm — Zero Adjustment Wheel

Pan →

Damper →

Balance Indicator

Before massing an object, always be sure the balance you are using is *zeroed*. This means that the two empty pans are in balance. If the balance is zeroed, the balance indicator will point to the exact center of its scale. If the indicator is off center, turn the *zero adjustment wheel* slightly until the indicator does point to the center of its scale. This zero adjustment wheel simply corrects for an out of balance condition. If the balance is not zeroed prior to massing, your results will not be accurate.

On the left side of the base is a small projection which on some balances resembles a paper clip. This is the *damper*. By pushing the damper toward the center of the balance you can reduce any up and down movement of the pans due to vibration or air currents in the room.

Stored at the base of the balance are the *standard gram masses*. These masses are the objects of known mass with which objects of unknown mass are compared. Simply place the object of unknown mass in one pan, then add masses to the other pan until the pans balance and the indicator points to the center. The mass of the unknown is the total of the masses it took to balance the pans.

And now a checklist for the procedure and some practice massing with an equal-arm balance. To mass an object, follow this procedure:

1. *Zero* the equal-arm balance; use the damper if necessary.
2. If you are massing out chemicals, *cover* the pan with a small piece of paper so that the balance is kept clean.
3. *Place the object* you wish to mass in the center of one of the pans.
4. *Add masses* to the center of the other pan until the balance arms are horizontal.
5. *Total the masses* it took to balance the unknown mass.
6 *Record* the observed mass.

⇨ *Obtain* from the supply area an equal-arm (double-pan) balance, 20 centicubes, a baby food jar, three large sinkers, and a set of masses. Find the mass of the objects as described in the chart below and record your finds in the column labeled *Your Masses*. In order that you may have a check for your measurements we have massed the same objects and recorded *Our Masses* in the chart. Slight differences may occur so you will have to exercise some judgement in comparing your masses with ours.

Object	Our Masses	Your Masses
5 centicubes	5 grams	
20 centicubes	20 grams	
baby food jar	84 grams	
three sinkers	85 grams	

Select three relatively small items in the lab. Pick each one up and try to estimate its mass. Then check your estimate using the equal-arm balance. Record the name of the object, your estimate, and the actual mass in the table below.

Object	Estimated Masses	Measured Mass
1.		
2.		
3.		

Are you thinking metric?!

Pull your thoughts about measuring metric masses together and fill in the following chart:

If you want to measure . . .	You should use a . . .	And measure in . . .
Mass of relatively large objects (about the size of a Hi C can or package of Velveeta cheese and larger)		
Mass of relatively small objects (between the size of a coin and the size of a coffee can)		
Very small masses (such as vitamins and pills)		

Compare your chart with the following one.

SELF-CHECK

If you want to measure . . .	You should use a . . .	And measure in . . .
Mass of relatively large objects (about the size of a Hi C can or package of Velveeta cheese and larger)	a scale or compare with known masses	kilograms (k)
Mass of relatively small objects (between the size of a coin and the size of a coffee can)	an equal-arm balance	grams (g)
Very small masses (such as vitamins and pills)	an equal-arm balance which is very sensitive	milligrams (mg)

A ctivity 5

MEASURING METRIC VOLUMES

The liter is the unit that is used with the metric system to measure how much liquid something can hold. The symbol *L* stands for liter.

⇨ *Go* to the Metric supply area and pick up a container labeled *L* liter. Fill the container with water and try to get an idea of how much 1 liter really is. ⇨ *Obtain* at least four other containers of different sizes and shapes, and try to estimate whether each would hold less than one liter, more than one liter, half a liter, two liters, three liters, etc. Then check your estimates by pouring the water from the liter-measure into each of the containers. Refill the container as needed. A liter is the amount of liquid that can be held in a container 1 dm by 1 dm by 1 dm (1 dm^3 = 1 liter).

Container	Estimate	Measure
1		
2		
3		
4		

Again, prefixes are used to show very large or very small quantities. Because most of the substances that you will be measuring are relatively small, you will need to learn about *milliliters*.

Pictured below is a centicube. A centicube measures 1 centimeter by 1 centimeter by 1 centimeter, so its volume is one cubic centimeter. If we filled this centicube with liquid, we would have *one milliliter* or liquid. Milliliter is symbolized *mL*.

Any container that is graduated in milliliters can be used to measure small amounts of liquid.

⇨ *Obtain* a graduated cylinder from the supply area and pour some water into it. If the graduated cylinder is glass, you should notice that the upper surface of the water is curved or crescent-shaped. This curved surface is called the meniscus. When you measure the volume of a liquid, you should line up your eyes, as shown in the diagram, with the bottom of the meniscus. If you are using a plastic graduated cylinder, it will not have a meniscus. What volume is shown in the above diagram?

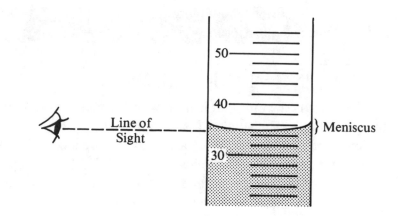

You should have read the volume as 35 mL.

As you have probably already learned, it is important to determine how many milliliters are represented by each mark on the graduated cylinder. It may differ from container to container. How much is contained in each of the following graduated cylinders?

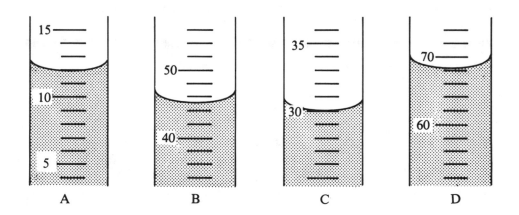

| A | B | C | D |

The answers are given below.

SELF-CHECK ✔

     a. 12 mL
     b. 45 mL
     c. 30 mL
     d. 68 mL

More about measuring volume. . .

The basic unit of measure for the volume of solids or space is the *cubic meter*. Your kitchen stove probably measures about 1 meter by 1 meter by 1 meter, or 1 cubic meter.

The *cubic centimeter* is used to measure the volume of small solid objects. Cubic centimeters is symbolized cm³.

Suppose you were asked to find the volume of a small rock. You would need to determine two things:

1. What *units* will you use in your answer?
2. What *process* will you use to do the measuring?

Since the object you are measuring here is solid and relatively small, the cubic centimeter is appropriate. If the rock is irregular in shape it would be impossible to measure its volume accurately with a ruler; but if you dropped the rock into a graduated cylinder of water, you would notice the water rising in the cylinder. If the water rose 4 mL, would the volume of the rock be 4 mL? Of course not, because your answer must be in cubic centimeters. You will have the right answer as soon as you find how a milliliter and a cubic centimeter compare. Do this simple activity to find out:

Use the centicubes to determine if 1 cubic centimeter is equal to 1 milliliter. (Remember a centicube is a cubic centimeter.) In the box below write the relationship between 1 cubic centimeter and 1 milliliter.

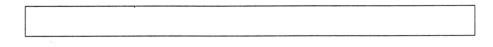

Now if you say that rock that displaced 4 mL of water has a volume of 4 cm³ you are right!

For practice . . .

⇨ *Obtain* a baby food jar, five marbles, four washers, and a graduated cylinder from the supply area. Find the following:

a. amount of water that a baby food jar holds
b. volume of five centicubes
c. volume of five marbles
d. volume of four washers

The answers are given below.

SELF-CHECK

a. baby food jar: 142 mL
b. five centicubes: 5 cm³
c. five marbles: 8 cm³
d. four washers: 2 cm³

Some of your results may be different depending on the actual materials used.

Gather your thoughts about metric volumes and fill in the following chart:

If you want to measure . . .	You should use a . . .	And measure in . . .
Volume of relatively large amounts of a liquid (such as large cans of paint, tanks of gasoline, cartons of milk)		
Volume of relatively small amounts of a liquid (as in tiny bottles of perfume, doses of cough medicine, small amounts called for in recipes)		
Volume of relatively large solids or amounts of space (such as the volume of a load of lumber, or the volume of a room)		
Volume of relatively small solids or amounts of space (such as the volume of a rock sample, or the volume of air in a small area)		

Compare your chart with the following one.

SELF-CHECK ✓

If you want to measure . . .	You should use a . . .	And measure in . . .
Volume of relatively large amounts of a liquid (such as large cans of paint, tanks of gasoline, cartons of milk)	liter measure	liters (L)
Volume of relatively small amounts of a liquid (as in tiny bottles of perfume, doses of cough medicine, small amounts called for in recipes)	cylinder, beaker, or measurer graduated in milliliters	milliliters (mL)
Volume of relatively large solids or amounts of space (such as the volume of a load of lumber, or the volume of a room)	meterstick or tape and measure length, height, width; $v = 1 \times h \times w$ or measure the amount of water the solid displaces	cubic meters (m^3)
Volume of relatively small solids or amounts of space (such as the volume of a rock sample, or the volume of air in a small area)	graduated measurer and measure the amount of water the solid displaces or if possible, measure length, height, width; $v = 1 \times w \times h$	cubic centimeters (cm^3)

Activity 6

MEASURING METRIC TEMPERATURE

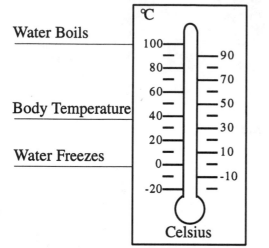

Water Boils

Body Temperature

Water Freezes

Celsius

Temperature in the metric system is measured with a Celsius thermometer. Examine the Celsius thermometer pictured here. Observe the three standard temperatures marked on the scale: the temperature at which water boils (100°), the temperature at which water freezes (0°), and normal body temperature (37°).

The symbol ° means *degree*. The temperature 20°C, for example, should be read as twenty degrees Celsius.

➪ **Obtain** a Celsius thermometer from the supply area. Examine the scale on the thermometer. Is the thermometer calibrated in one degree intervals? Two degree intervals? Five degree intervals? Measure the room temperature.

1. Number of degrees per interval on the scale _____ .
2. Room temperature is _____ .

Check your answers with someone else's.

➪ **Obtain** (in addition to the Celsius thermometer) a baby food jar or other container and some ice from the supply area. Fill the container half full of cold water. Measure the temperature of the water. Be sure to give the thermometer a few seconds to adjust before reading it. Add an ice cube to the water. Measure the temperature every two minutes while stirring the water gently. Complete the following table:

Time	Temperature
Temperature of water before adding ice	_____
Temperature of water after adding the ice	_____
After 2 minutes	_____
4 minutes.	_____
6 minutes	_____
8 minutes	_____

Compare your answers with the ones that follow.

SELF-CHECK ✔

Your table should be like this:

Time	Temperature
Temperature of water before adding ice	18°C
Temperature of water after adding the ice	
After 2 minutes	13°C
4 minutes	12°C
6 minutes	10°C
8 minutes	10°C

Don't be disturbed if your answers aren't the same as ours. A lot of factors can influence the results. The important thing is that you should be able to read the Celsius thermometer.

Optional Activity for Measuring

MEASURING FORCES

Whenever you measure a push or pull, you are measuring force. The unit of force in the metric system is the *Newton* (N). Newtons are a measure of how much force is being exerted on an object.

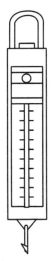

Instruments like the spring scale shown on the left and the personal bathroom scale are used to measure force. The spring scale may be hung, held, or laid on a flat surface. The greater the force being measured, the farther the spring is stretched.

A simple force measurement for measuring small forces could be made using a rubber band. Hang objects on the rubber band and calibrate the distance each stretches the band.

Weight is a force. It is the pull of gravity on nearby objects. The earth's gravitational pull on nearby objects is stronger than the moon's gravitational pull on objects near its surface. A person would weigh about one sixth of his earth weight on the moon.

Keep in mind that mass is the measure of the amount of matter in an object. Therefore, the mass of an object would remain the same no matter where in the universe it is placed.

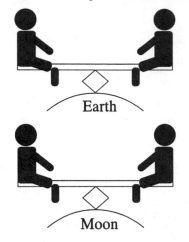

Earth

Moon

Two children balancing one another on a teeter-totter on earth would still balance one another on the moon. Each child's *mass* would be the same on the moon as on the earth, but each child would *weigh* about one-sixth as much on the moon as on earth.

If we keep our thoughts earth bound, we can state a definite relationship between mass and weight. At sea level, on the earth's surface, a kilogram mass weighs about 10 newtons.

Continue on and take the following Self-assessment.

Self Assessment: Measuring Metrically

I. In the space provided, write the most appropriate metric unit in which each of the following should be measured:

1. The length of a basketball court should be measured in _____ .

2. The temperature of water should be measured in _____ .

3. The volume of an orange should be measured in _____ .

4. The diameter of a pencil should be measured in _____ .

5. The mass of a refrigerator should be measured in _____ .

6. The width of a VCR should be measured in _____ .

7. The amount of water used in a baking recipe should be measured in

_____ .

8. The amount of gas an automobile gas tank can hold should be measured in

_____ .

9. The mass of a package of chewing gum should be measured in

_____ .

10. The mass of a postage stamp should be measured in _____ .

II. Use what you know about metric prefixes and state equivalent metric measures for each of the following:
   1. 1 meter = _____ decimeters

   2. 1 dm = _____ cm = _____ mm

   3. 1 mm = _____ m = _____ dm = _____ cm

   4. 2 kilometers = _____ meters

   5. 4000 m = _____ km

   6. 500 meters = _____ hectometers

   7. 1 km = _____ dam

   8. 1 km = _____ hm

   9. 1 liter = _____ milliliters

III. Take a *Measurement Test Packet* from the supply area and —
   1. Measure the length and width of the packet envelope to the nearest 0.5 centimeters.
   2. Remove a carpenters' nail from the envelope and measure its length to the nearest millimeter.
   3. Remove ten buttons from the envelope and measure their mass (weight) to the nearest gram.
   4. Find the volume of the buttons.
   5. Record the present room temperature in degree Celsius.
   6. Measure the length of the line drawn below. Give your answer in millimeters, centimeters, decimeters, and meters.

   _____

The answers to the self-assessment follow.

SELF-ASSESSMENT ANSWERS

I.
1. length of basketball court—meters (m)
2. temperature of water—degrees Celsius (°C)
3. volume of an orange—cubic centimeters (cm^2)
4. diameter of a pencil—millimeters (mm)
5. mass (weight) of a refrigerator—kilograms (kg)
6. width of a VCR— centimeters (cm)
7. amount of water used in a recipe—milliliters (mL)
8. amount of gas a tank can hold—liters (L)
9. mass (weight) of a package of gum—grams (g)
10. mass (weight) of a postage stamp—milligrams (mg)

II. 1. 1 meter = <u>10 decimeters</u>
   2. 1 dm = <u>10 cm</u> = <u>100 mm</u>
   3. 1 mm = 1/1000 m = <u>1/100 dm</u> = <u>1/10 cm</u>
   4. 2 kilometers = <u>2000 meters</u>
   5. 4000 m = <u>4 km</u>
   6. 500 meters = <u>5 hectometers</u>
   7. 1 km = <u>100 dam</u>
   8. 1 km = <u>10 hm</u>
   9. 1 liter = <u>1000 milliliters</u>

III. 1-5. Check your answers with those given on the card in the *Measurement Test Packet*.
   6. length of line: 115 mm, 11.5 cm, 1.15 dm, .115 m

IDEAS FOR YOUR CLASSROOM

Measuring for the sake of measuring is dull and pointless, so incorporate measuring with other class activities. Here are some suggestions which you can expand with your own ideas. The activities you choose should depend a great deal on your students' interests.

1. Construct other measuring instruments from the ones already available in the classroom. What would you need to measure the distance between the office and the cafeteria? The distance around your waist? The distance from the second story window to the ground?
2. Measure growing plants. Keep a record and make comparisons between plants grown in different conditions.
3. Construct a map of the classroom representing actual distances.
4. Measure shadows at different times of day.
5. Keep a record of the amount of food and water a classroom pet requires each day.
6. Here is a (metric) recipe for growing crystals. The results are fascinating!
5 mL household ammonia
15 mL water
15 mL table salt
15 mL bluing
Mix together and pour over rocks, sand, sponges, wood, or bits of brick and cement. Spread out on a metal tray. Let it stand and watch the crystals grow.
7. Make bread, cookies, or pudding. Measure the ingredients, temperature, and time.

THOUGHT-STARTER QUESTIONS

- Before you measure, what is your estimate?
- What is this object's length, mass weight, volume, area, and so on?
- Quantify that observation.
- Use numbers to describe what you observe about this object or event.
- Measure and record what you observe about this object or event?
- How long (heavy) do you think this object is?

ASSESSING FOR SUCCESS: PERFORMANCE TASK

Preparation

You will need to set up labeled stations around the room with the materials listed below for each station. Manage the flow of students by having some students start at different stations. If the room size allows, have more than one set of stations.

Directions to the Students

1. Go to each station set up in the room (as directed by your teacher). For each station follow the directions given on this page.
2. Answer the questions for each station in the space provided on this page.
3. When recording your measurements, record both the number and the unit of measurement (for example, 10 centimeters).

Station A

Check your materials: metric ruler and meter stick.

_____ 1. Measure the width of this paper in centimeters.
_____ 2. Measure the length of this line in millimeters.
_____

_____ 3. Measure the length of this tabletop in meters.

Station B

Check your materials: graduated cylinder, plastic cup with a black line, container of water.
_____ 4. Pour water into the cup until it reaches the black line. How much water is in the cup?

When you are finished, pour the water back into the container.

Station C

Check your materials: dual scale thermometer.

_____ 5. What is the room temperature in degrees Celsius?

Station D

Check your materials: two-pan balance, set of masses, object to be massed

_____ 6. What is the mass of the object?

When you are finished, remove the masses and the object from the balance.

Inferring

To do the inference activities you will need:

- ✓ coin
- ✓ pencil
- ✓ a plastic sandwich bag
- ✓ a water source
- ✓ 2 magnetic compasses
- ✓ magnet
- ✓ C-sized battery
- ✓ 50 cm length of insulated copper wire, ends stripped
- ✓ a mystery box

WHY IS INFERRING IMPORTANT?

We have a better appreciation of our environment when we are able to interpret and explain things happening around us. We learn to recognize patterns and expect these patterns to reoccur under the same conditions. Much of our own behavior is based on the inferences we make about events. Scientists form hypotheses based on the inferences they make regarding investigations. As teachers we constantly make inferences about why our students behave as they do. Learning itself is an inference made from observed changes in learned behavior.

GOALS

In these exercises you will learn about model building and develop skills necessary to make proper inferences based on observation.

PERFORMANCE OBJECTIVES

After completing this set of activities you should be able to:

1. Given an object or event, construct a set of inferences from your observations about that object or event.
2. Given additional observations about the object or event, identify the inferences that should be accepted, modified, or rejected.

CONSTRUCTING INFERENCES FROM OBSERVATIONS

While an observation is an experience perceived through one or more of the senses, an inference is an *explanation or interpretation of an observation*. Suppose, for example, you look out your window and see two men carrying a TV set away from your neighbor's house. What you actually observe is the men carrying the TV. You might wonder and even attempt to explain why the men are carrying the TV. There may be several reasons why men would carry a TV set away from a house. Perhaps . . .

- Someone bought the set from the neighbor and is taking it to their own home.
- The TV is being picked up to be repaired.
- The owner bought a new set and is trading this one in.
- The set is broken and is being discarded.
- The TV is being stolen.

Perhaps you can think of some explanations for what was observed other than those mentioned. Each of the statements that were used to logically explain what as observed is called an *inference*.

We use our past experiences to build mental models of how the world works. New experiences only make sense to us when we are able to link them to understandings we already have.

To *infer* means to construct a link between what is observed directly and what is already known from past experience. A *map* of the process of inferring might look like this.

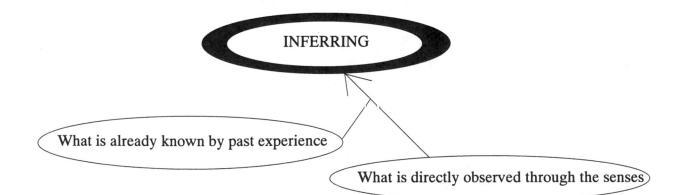

An inference is a statement that goes beyond the evidence and attempts to interpret or explain a set of observations. It follows that every inference must be based on observation. An inference is NOT a *guess* since a guess is an opinion formed from little or no evidence: Here are some examples of observations followed by inference statements:

- The brass knob on that door is not bright and shiny. I infer that the office is not used often.
- There is a spot in my front yard where grass does not grow. Someone may have spilled a toxic substance there.
- I see that iodine turns purple when I put it on a potato chip. It can be inferred that the chip has starch in it.
- The pages in this book are yellow. I infer either that the book is old or that the paper was dyed yellow to give it an old appearance.
- Through the window I see the flag waving. It must be windy out.
- The fish are floating on top of the tank. Perhaps no one fed the fish.
- My drinking water smells like rotten eggs. Maybe it has become contaminated.
- The cabbages that were growing in my garden are gone and there are droppings on the ground. That is evidence that rabbits have been there.
- That star is brighter than the others. I infer it is closer to Earth than the others.

Note that each inference was based on observation. Also note that it is often easy to tell which statements are observations and which are inferences simply by the way they are phrased.

When inferring, it is helpful to follow these steps:

1. Make as many observations about the object or event as possible.
2. Recall from your experiences as much relevant information about the object or event as you can and integrate that information with what you observe.
3. State each inference in such a way that clearly distinguishes it from other kinds of statements (observation or prediction):
   "From what I observe I infer that . . ."
   "From those observations it can be inferred that . . ."
   "The evidence suggests that . . . may have happened.
   "What I observe may have been caused by . . ."
   "A possible explanation for what I see is that . . ."
   "From what I observe I conclude that . . ."

In this first activity you will be constructing some inferences of your own. Remember that an observation is an experience perceived through one or more of the senses. An inference is an explanation or interpretation of an observation.

Activity 1

CONSTRUCTING INFERENCES FROM OBSERVATIONS

Take out a coin . . . any coin . . . and examine it carefully. Do the following:

1. Make as many observations as you can about the coin (remember to describe properties) and list them in the observation column below.
2. Try to think of things you already know that might help to explain or interpret what you observe about the coin.
3. In the inference column list as many inferences as you can about your coin, being sure to state each inference in such a way that indicates that it is an inference.
4. Draw a line between each inference and the observation on which it is based. (More than one inference may be drawn from one observation.)

Just to help you get started, here are a couple of possible observations and inferences someone else might make about his or her coin:

Observations	Inferences
The surface is dark and dull rather than shiny and bright.	The coin has been handled a lot. The coin has rubbed up against other coins.
The date 1994 appears on one side.	I infer that the coin was made in 1994

COIN	
Observations	**Inferences**

SELF-CHECK ✓

The following are some observation and inference statements that someone else made about his or her coin. Because your coin and your own past experiences are different, your observations and inferences will differ. The Self-Check will help you to decide if you applied the skills of observing and inferring properly. You may have phrased your inferences slightly differently.

COIN	
Observations	**Inferences**
This coin is the color of copper.	I infer it is made of copper.
This coin has the date 1994 marked on it.	The coin probably was made in 1994
This coin has raised letters on it and they are clear and uniform in size.	I infer the coin was made by machine.
The coin has a green substance on one side.	Perhaps the coin sat in water and became corroded.
When I drop the coin on the table it makes a "clinking" sound.	I infer the coin is solid rather than hollow.
There are a lot of little short scratches on both sides and edges of the coin.	I infer the coin has been carried in someone's pocket or purse with other coins.
The coin has one long deep scratch on one side.	Maybe someone deliberately gouged the coin with a sharp instrument.

Activity 2

FORMULATING INFERENCES FROM OBSERVATIONS ABOUT EVENTS

This next activity is designed to help you learn to formulate inferences about events. Every inference must be based on an observation, so you will first be making careful observations and then interpreting or explaining those observations. These interpretations or explanations of observations are inferences. (This is also a very good activity to use as inferential thinking in creative writing!)

Observe these *tracks in the snow*. To help you think more logically about the picture, it has been separated into frames. Make at least two observations about each frame, and for each observation write at least one inference that could be drawn from that observation. (More than one inference can be drawn from one observation.) Draw a line from each inference to the observation on which it is based.

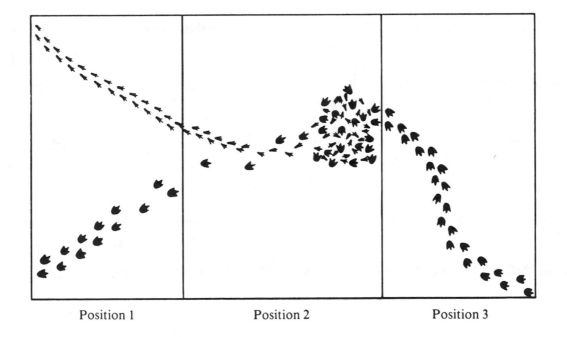

Position 1   Position 2   Position 3

Observations	Inferences
Example: Large footprints get farther apart. — The animal is stepping over stones. The animal is running.	
Position 1	
Position 2	
Position 3	

Check your inferences with someone else's or with those listed in the Self-Check.

SELF-CHECK ✔

Here are just a few inferences you might have made. There could be many, many more. In checking your inferences be certain that each is based on specific observations. More than one inference can be drawn from a single observation.

Observations	Inferences
Example: Large footprints get farther apart.	The animal is stepping over stones. The animal is running.

Position 1

Observations	Inferences
One set of prints is smaller than the other.	One animal is smaller than the other.
The small and large prints are headed in the same direction.	Both are walking toward something.
There are three toes for each print.	Both animals are birds.
The small and large prints get closer to each other.	The large animal is pursuing the small animal. Both are walking in a gully.
The larger prints get farther apart.	The larger bird is going downhill. The larger bird is running. The larger bird is stepping over stones.

Position 2

Observations	Inferences
The prints converge.	The larger animal catches and either eats or carries off the smaller animal. The animals were there at different times. Both animals discovered food in the same place.
The prints become all mixed up.	The animals were milling about. The animals were fighting.
The smaller footprints get farther apart.	The smaller animal begins to run.

Position 3

Observations	Inferences
The small tracks stop.	The larger animal ate the smaller one. The smaller animal flew away. The snow at this point became crusty and the smaller animal was not heavy enough to make tracks.
The large prints are close together.	The larger animal was walking rather than running.

LEARNING IS AN INFERENCE

Recall that an inference is an explanation or interpretation of what you observe. When you are inferring you try to give meaning to what you observe. Learning . . . making sense of things . . . then, is an inference.

Because inferences are based not just on what is observed but also on what the observer already knows, it follows that new experiences may be interpreted differently by different people. Each person, constructs his or her own learning depending upon past experiences and knowledge already gained. Not everyone walks away from the same experience having constructed the same new knowledge. Therein lies a problem, particularly for teachers. For children to learn something new they must relate the new concept to concepts they have already formed. If an individual child is unable to relate the new concept to any pattern of knowledge he or she already has, learning does not take place. The teacher's role is to facilitate learning by linking new concepts with what individual children already know.

You, as a teacher, must be an expert observer and questioner. In order to link new knowledge to old knowledge, you will need to find out what that previous knowledge is. You can not assume it exists for all students. By creating the right situations, you can observe what individual children know and do not know. When prerequisite knowledge does not exist, your role as a teacher is to design activities in which children can use their senses to experience as much about a concept as they can. That experience base will in turn allow them to *construct* more knowledge.

CONFIDENCE IN INFERENCES

Patterns develop from similar or related past experience. Observations that can be linked to patterns of known information form a basis for making inferences. If you observe something new that matches a pattern of information you already have, then you have considerable confidence that your inference really does explain what you observed. On the other hand, you may have little confidence in your inferences when observations do not match a preexisting pattern.

Just as in previous activities, in the following activity you will make both observations and inferences. This activity, however, differs in that you will also be asked to indicate how confident you are in what you infer. When making your inferences, be sure to use only words and ideas you understand. Your confidence in some inferences about this activity is likely to be high when your observations fit into a pattern of experience you already possess. When observations do not match a previously learned pattern very well, your confidence in accurately explaining what you observe is apt to be much less.

Activity 3

This activity involves the use of water. Do the activity over a sink or a container that can catch any spilled water.

⇨ *Go* to the supply area and obtain a new well sharpened pencil. List at least *five* observations and inferences about the pencil. Each time you make an inference draw a line between that inference and the observation on which you based it.

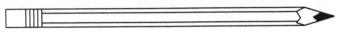

Pencil	
Observations	**Inferences**
1.	
2.	
3.	
4.	
5.	

Look at your list of inferences and (circle) the ones in which you have a great deal of confidence.

⇨ *Obtain* a plastic sandwich bag from the supply area. Fill the bag about two thirds full with water leaving enough room for you to grasp the bag and hold it closed with one hand. The bag should not leak. List at least *five* observations about the bag and its contents.

Bag of Water	
Observations	**Inferences**
1	
2.	
3.	
4.	
5.	

Look at your list of inferences and (circle) the ones in which you have a great deal of confidence.

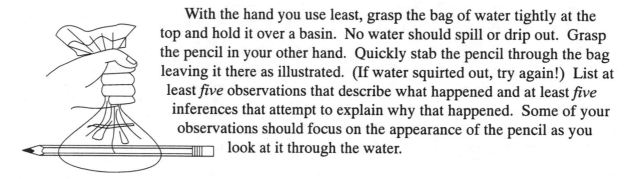

With the hand you use least, grasp the bag of water tightly at the top and hold it over a basin. No water should spill or drip out. Grasp the pencil in your other hand. Quickly stab the pencil through the bag leaving it there as illustrated. (If water squirted out, try again!) List at least *five* observations that describe what happened and at least *five* inferences that attempt to explain why that happened. Some of your observations should focus on the appearance of the pencil as you look at it through the water.

Pencil and Bag	
Observations	Inferences
1.	
2.	
3.	
4.	
5.	

Look at your list of inferences and (circle) the ones in which you have a great deal of confidence.

SELF-CHECK ✓

You probably have confidence in the inferences you made concerning what the pencil and bag might have been made of, where they came from, what they contain and what they were made for. You have observed and experienced these objects before.

You may have less confidence, however, in the inferences you made to try to explain what you observed as the pencil was jabbed into the bag. Your lack of confidence in explaining why water did not spurt out and why the pencil appears bent is due to never having observed and interpreted this in the past.

ACCEPTING, REJECTING, AND MODIFYING INFERENCES

Often, after having drawn inferences from a set of observations, new information becomes available that may cause you to rethink your original inferences. Sometimes additional observations reinforce your inferences. At other times, however, additional information may cause you to modify or even reject inferences that were once thought to be useful. New observations lead to adjusting patterns of experience to accommodate the new information. The science processes of observation and inference might look something like this.

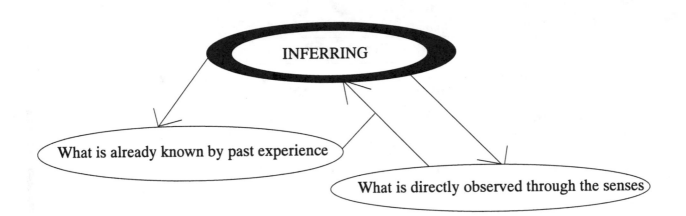

In science, inferences about how things work are continuously constructed, modified, and even rejected on the basis of new observations.

In the following activity you will make some initial observations about an event and draw some inferences from those observations. You will then have the opportunity to make additional observations. Your new observations will cause you to either accept, modify, or reject your first set of inferences. When you have finished, you will be asked to write one more inference, called a *conclusion*, which is a summary of all the inferences you decided to accept.

Activity 4

⇨ *Go* to the supply area and pick up a magnetic compass. (You will later need a magnet, another magnetic compass, a C-sized battery, and an insulated copper wire with ends stripped, but do not get these other items now.)

Set the compass on the table in front of you and move it around a bit making sure there are no other compasses or magnets in the immediate area. Record what you observe about the behavior of the compass needle and where it points. Write down as many inferences as you can about what you observe.

Observations	Inferences

⇨ *Obtain* a magnet and bring it closer to your compass. Move it around the compass. Record what you observe happening. Write as many inferences as you can to explain what you observe.

Observations	Inferences

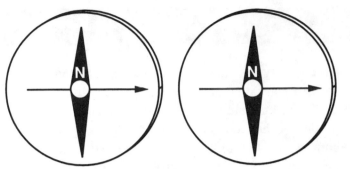

Remove the magnet from the area where your first compass is located. Obtain another compass and place it near the first compass. Move them both around. What happens? Record what you observe and again make as many inferences as you can about what you observe.

Observations	Inferences

⇨ *Go* to the supply area and obtain a battery and a length of insulated copper wire (ends stripped) that is about 50 cm long. Place the compass away from the battery so it does not react with the steel in the case. Hold one end of the wire to one of the battery terminals. Place the compass next to the wire as illustrated below. Intermittently touch the other end of the wire to the other battery terminal.

⇨ *Observe* what happens to the needle of the compass as you open and close the circuit. List your observations and possible inferences below.

Observations	Inferences

DRAWING CONCLUSIONS

This series of activities related to magnetism may have reinforced what you already knew about magnetism or may have caused you to adjust your thinking (patterning) about what magnetism is. After all you can't see magnetism—you can only see what it does. To find out what adjustments, if any, you have made in your thinking about magnetism, do the following:

1. *Review* all the inferences you have made in this entire activity and accept, modify, or reject them by following these steps:

   - If you made inferences that you think are no longer good explanations for what you observed, reject them by drawing a line through them.
   - If you made inferences that you think are still OK but need some slight modifications to adequately explain what you observed, change them.
   - If you made inferences that you still feel are good explanations for what you observed, *circle* them to show that you accept them.

2. *Draw a conclusion:*

   Write a statement below that *summarizes* what you were able to infer from all of your observations. Make your inference statements simple and use only words that are meaningful to you. You are trying to explain phenomena in the context of what you already know.

   From my activities about magnetism I conclude that:

   _____

   _____

   _____

   _____

   _____

Of course, what you concluded will always be subject to revision and modification as your experience grows.

SELF-CHECK

Here are some conclusions other students have drawn from doing the activities related to magnetism.

- *Something* is present in the environment that has the same effect on a compass that a magnet does.
- The Earth behaves as if it were a giant magnet.
- There is a relationship between magnetism and electricity.

Self Assessment Inferring

⇨ **Go** to the inference supply area and pick up a mystery box. Do not open the box or in any way tamper directly with the contents. Make at least five observations and five inferences about the contents of the box. Identify the specific observation on which you base each inference by drawing a line between them. You may tilt, shake, roll, or rattle the box but do not peek inside. List your observations and inferences in the chart below.

Observations	Inferences
1.	
2.	
3.	
4.	
5.	

Now open one end of the box and without looking inside put your hand in the box and gather some additional information about the contents. In the chart below, accept, reject, or modify each of your original inferences on the basis of this new information. Identify the observations on which you accept, reject, or modify each inference.

Observations	Inferences
1.	
2.	
3.	
4	
5.	

Compare your observations and inferences with your partner's or check your answers with those in the envelope taped to the mystery box.

IDEAS FOR YOUR CLASSROOM

1. *Pictures* are excellent for use in developing skills of observation and inference. Use pictures showing action that has already taken place and have students make observations and inferences about their observations. Comic strips, cartoons, coloring books, and comic books are good sources of pictures. Pictures of animals are also excellent for developing observation and inference skills. Organisms are adapted for their survival (e.g. coloration for protection, feet adapted for catching prey, feet for escaping predators, spines on cacti for protection, color to attract animals to eat the fruit like grapes, etc.)

2. *Mystery Boxes* are fun and intriguing as well as excellent activities for observations and inference making. Enclose unknown objects in a shoe box and have the students make as many observations as possible without opening the box. Try to involve as many senses as you can except sight, by providing the means to the student to feel or smell the object. Give students practice in accepting, modifying or rejecting inferences on the basis of additional information. Some objects that could be used include a sugar cube, bar of soap, a toothbrush, pine cone, popcorn, lemon, onion, or a stick of gum. There are many other objects that could be used.

3. *Unknown Gases* may be best done as demonstrations but students can participate. These exercises are wonderful for having students make inferences.

Oxygen

First you'll need:

- ✓ glass jar or clear plastic container
- ✓ enough 3% hydrogen peroxide to fill the jar about half full
- ✓ cake or package of yeast
- ✓ candle (or splint) and matches

1. Fill the jar about half full with hydrogen peroxide and sprinkle some yeast into the peroxide. Have students record their observations. (What they *don't* see, smell, taste, feel, or hear can be an observation too.)

2. Add a little more yeast, then light a candle or splint and lower the flame into the jar above the liquid. What happens?

3. Blow out the candle and lower the glowing wick into the jar. What happens? What inferences can be made from your observations? (This is really an oxygen generator but students can only *infer* the presence of oxygen as they observe the burning candle flame up and burn faster and the glowing wick burst into flame again.)

Carbon Dioxide

For this activity you'll need:

- ✓ baking soda
- ✓ vinegar
- ✓ candle
- ✓ matches
- ✓ plastic shoe box (or some small container)
- ✓ peanut butter jar (or similar size jar)
- ✓ piece of clay

1. Pour about a cup of baking soda into the plastic box.
2. Pour about 2 or 3 cups of vinegar into the soda in the shoe box. What do you observe?
3. Lower a burning match into the shoe box just above the solution. What happens? Do it again. (Students should see that the flame goes out. They might infer that carbon dioxide has been generated and that explains why the flame goes out.)
4. Secure the candle with the clay to the bottom of the inside of the peanut butter jar. Then light the candle.
5. Carefully *pour* the carbon dioxide gas over the burning candle. What happens? Why?

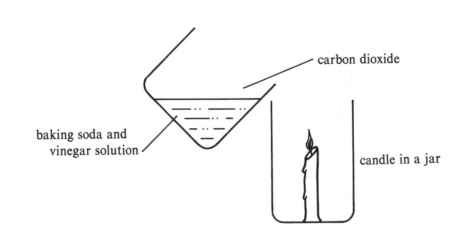

baking soda and vinegar solution

carbon dioxide

candle in a jar

ASSESSING FOR SUCCESS: PERFORMANCE TASK[1]

Preparation

One way to assess a student's ability to construct inferences based on observations is to construct mystery boxes. Be sure to use identical boxes and objects when assessing children simultaneously. An example of a mystery box is one that uses two small objects that either roll or slide. Inexpensive and readily available rolling objects include BB's, ball bearings, plastic Easter eggs, and marbles. Metal washers and bottle caps and lids are good sliding objects.

Directions to the Student

1. Write your name on this line _____ .
2. Check the box to be sure it is taped shut. Do not open the box.
3. The box contains one or more objects. Pick up the box and listen to the sounds as you gently shake it. Tilt the box and listen carefully.
4. Answer the following questions:

_____ a. What shape is an object in the box?

_____ b. What is one other property of an object in the box?

_____ c. What is one kind of motion made by an object in the box?

_____ d. Except for air, how many objects do you think are in the box?

_____ e. Explain why you think your answer to question d is correct?

Scoring Procedure

1 point each for questions a, b, c, and e. Question d is not scored separately.

Acceptable responses for questions a, b, c, and e include:

a. flat, round, ball-shaped, like a coin, and so on.
b. hard, heavy, sounds like metal.
c. slides, rolls, glides, or drops.
e. student response has to support response to question d.

1. Adapted from an example presented by Douglas Reynolds at The United States Department of Education Secretary's Conference on Improving Assessment in Mathematics and Science Education. September 20-21, 1993, Arlington, Virginia.

THOUGHT-STARTER QUESTIONS

- What are some possible reasons why that happened?
- What are some other explanations for what you observed?
- What might have caused that to occur?
- What is a logical explanation for that?
- In your own words explain why that happened.
- How do you interpret this evidence?
- What does this set of observations mean to you?

Predicting

Sight
Smell
Sound
Taste
Touch

To do the *Predicting* activities that follow you will need:

- ✓ a pendulum support (nothing special, even a desk will do)
- ✓ a string or cord
- ✓ 2 pendulum bobs (sinkers, washers, or any small weights)
- ✓ a container of buttons (25 red, 10 blue, 10 green, and 5 white)
- ✓ a meter stick

For the optional activity you will need:

- ✓ a glass jar (a liter or larger)
- ✓ enough particles to fill the jar (peas, marbles, rice, and so on.)

WHY IS PREDICTING IMPORTANT?

A prediction is a forecast of what a future observation might be. The ability to construct dependable predictions about objects and events allows us to determine appropriate behavior toward our environment. Predicting is closely related to observing, inferring, and classifying; an excellent example of a skill in one process being dependent on the skills acquired in other processes. Prediction is based on careful observation and the inferences made about relationships between observed events. Remember that inferences are explanations or interpretations of observations and that inferences are supported by observations. Classification is employed when we identify observed similarities or differences to impart order to objects and events. Order in our environment permits us to recognize patterns and to predict from the patterns what future observations will be.

Children need to learn to ask such questions as *If this happens, what will follow? What will happen if I do this?* As teachers, we need to be very careful about the kinds of predictions we make about student behavior and performance.

GOALS

In these exercises you will learn to construct predictions based on patterns of observed evidence and to test your prediction for dependability. New observations can be used to revise both predictions and inferences.

PERFORMANCE OBJECTIVES

After completing this set of activities you should be able to:

1. Distinguish among observation, inference, and prediction.
2. Construct predictions based on observed patterns of evidence.
3. Construct tests for predictions.
4. Use new observations to revise predictions and inferences.

Activity 1

DISTINGUISHING AMONG OBSERVATION, INFERENCE, AND PREDICTION

The following brief definitions may help you distinguish among observation, inference, and prediction.

- Information gained through the senses: *Observation*
- Why it happened: *Inference*
- What I expect to observe in the future: *Prediction*

The following activity is intended to give you practice in distinguishing among these important processes. Read the first two frames of the cartoon and the statements that follow. Indicate whether each statement is an observation, inference, or prediction. (Take the point of view of the cartoon characters.)

B. C. by permission of Johnny Hart and Field Enterprises, Inc.

1. *In about 2 minutes that mountain is going to blow sky-high.* _____

2. *I can feel the rumbling (earth vibrating) beneath my feet.* _____

3. The *rumbling* is caused by the volcano. _____

Was the prediction based on careful and comprehensive observation? How much confidence do you have in this prediction? To see how the cartoon turns out, look below.

SELF-CHECK ✓

Compare your answers with someone else's or check your answers with those below.

1. <u>Prediction</u> (A forecast of what a future observation will be.)
2. <u>Observation</u> (Information gained through the senses.)
3. <u>Inference</u> (An explanation for the observation.)

B. C. by permission of Johnny Hart and Field Enterprises, Inc.

The process skills of observing, inferring, and predicting can be clearly defined and each is clearly distinguishable from the others. You will see later that there is also a great deal of interdependence among these processes.

We make sense of the world around us by observing things happen and then interpreting and explaining them. We often detect patterns in what we observe. When we think we can explain why things work the way they do, we construct mental models in our heads that at least temporarily serve to provide order to things. Often we use these mental models to predict occurrences that might happen in the future. Here are some examples of predictions:

- *I see it is raining and the sun is coming out. There could be a rainbow.*
- *When I flip the switch the lamp will light.*
- *The weak magnet picked up five paper clips; I predict the strong magnet will pick up more.*
- *If I release both balls at the same time, they will hit the ground at the same time.*

Notice that each of the sample predictions is written in future tense. Each prediction statement is based on observations and patterns that have developed from past observations. How we explain and how we interpret what we observe affect how we predict.

A *map* of the process of predicting might look something like this:

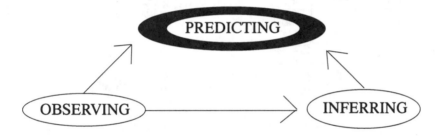

Predictions are *reasoned statements* based not only on what we observe but also on the mental models we have constructed to explain what we observe. Predictions are not just wild guesses because guessing is often based on little or no evidence.

In order to use the process skills of observing, inferring and predicting correctly you need to be able to clearly distinguish among them. The first activity was intended to provide you with some brief working definitions and to give you practice in distinguishing among observing, inferring and predicting. (You may want to refer to the separate learning activities for a thorough treatment of each process skill.)

TESTING OUR MENTAL MODELS

At one time in history many people believed that the Earth was flat. The Earth *looked* flat; that observation as well as others seemed to support the inference that the Earth was flat. People predicted that if sailors traveled far enough, their ship would fall off the Earth. Many people had a great deal of confidence in that prediction. Later, when sailors tested that prediction, they observed that their ship did *not* fall off the Earth. New observations caused people to change both their inference about how the Earth was shaped and their prediction about falling off the Earth.

Testing our predictions leads to making more observations that either support or do not support original predictions. When new observations are consistent with our predicted pattern of observations, we have even greater confidence in our prediction. However, when new observations do not support our original prediction, we may reject it and re-examine our observations. New observations lead to new inferences and new predictions. Therefore, our *map* of the process skill of predicting looks more like this:

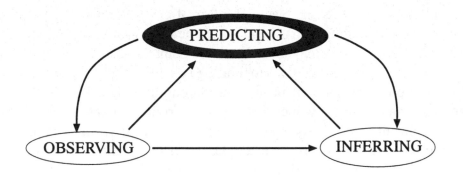

As new data (observations) are collected, theories (inferences) are proposed to explain what has been observed and to predict what has not yet been observed. In fact, for a theory to be accepted in science, it must meet a threefold test:

1. can explain what has already been observed
2. can predict what has not yet been observed
3. can be tested by further observation and modified as required by the new data.

Observing, inferring, and predicting are interconnected thinking skills. We use these skills to make sense of our world. Our ideas of how things work should be under constant review and subject to revision. Science should always be viewed as tentative; always subject to change as new observations result from testing our predictions.

Activity 2

CONSTRUCTING PREDICTIONS BASED ON OBSERVED PATTERNS

Predicting is stating what you think a future observation will be. You may recall that observations may be qualitative or quantitative in nature. While the activities you will do next involve both kinds of observations (qualitative and quantitative) you will be using some special tools to help manage the numbers in the quantitative observations you make.

Numbers are like any other bits and pieces of information; if you leave them lying around loose they have very little meaning. Charts and graphs help to organize information for easy review and retrieval. Organizing numbers in charts and graphs is also an effective way to communicate information to other people. Hopefully you have noticed the use of several *concept maps* in this book which are intended to guide your thinking and to provide a framework into which you can classify new ideas. It may be helpful for you to think of charts and graphs as kinds of concept maps. They are particularly useful in showing the relationships between two or more ideas. This chart, for example, shows the relationship between the date and the time of day the sun appears to rise at a certain place on Earth.

Date	Sunrise Time	Date	Sunrise Time
January 1	7:24	May 1	5:00
January 15	7:20	June 1	4:31
February 1	7:12	July 1	4:33
February 15	6:52	August 1	4:56
March 1	6:35	September 1	5:25
March 15	6:08	October 1	5:54
April 1	5:42	November 1	6:28
April 15	5:21	December 1	7:01

Please be aware that *sunrise* is really a misnomer since the sun does not rise and set. It is really the Earth turning on its axis that causes the sun to appear to rise and set.

But what about days in the year not shown in the table? Would it be possible to predict *sunrise* times for those days *not* directly observed? Let's see . . .

We make predictions by first looking for patterns. Answer the following questions designed to help you find a pattern in the observed *sunrise* times:

1. What time did the sun appear to rise on Jan. 1? _____ On Feb. 1? _____
2. Would you expect the *sunrise* time for Jan. 15 to be about halfway between *sunrise* times for Jan. 1 and Feb. 1? _____ Is it? _____ (Check the observed time.) (Note that Jan. 15 is not exactly *halfway* between Jan. 1 and Feb. 1 but it is close.)

3. Use the *halfway* method to predict the *sunrise* time for Feb. 15. (Try not to look until you have figured it.) What is your Prediction? _____ Then check your prediction with the observed *sunrise* time in the table.

4. Try this one . . . predict the sunrise time for October 15. _____ Then check the answer in the Self-check.

5. Suppose you wanted to predict a sunrise time for a date that was not halfway between two other given dates? Using the table, determine the sunrise time for September 10 and then check your answer in the self-check.

SELF-CHECK ✓

For # 4. If you used the *halfway* method to find the predicted sunrise time for October 15, your calculations probably look something like this:

```
    November 1     6:28
  - October 1      5:54
                     34   minutes difference
```
Half of 34 minutes is 17 minutes.

```
    October 1      5:54
  +                  17
                   6:11   predicted sunrise time
```

If you used the more exact method of calculating, your figures might look like this:

```
    November 1     6:28
  - October 1      5:54
                     34   minutes difference
```

14/31 of 34 minutes is about 15 minutes.

```
    October 1      5:54
  +                  15
                   6:09   predicted sunrise time
```

For # 5. September 10 is about one-third the way between September 1 and October 1.

```
    October 1      5:54
  - September 1    5:25
                     29   minutes difference
```

10/30 or 1/3 of 29 minutes is about 10 minutes.

```
    September 1    5:25
  +                  10
                   5:35   predicted sunrise time for September 10
```

A ctivity **3**

PRACTICE IN MAKING PREDICTIONS

When making predictions it is important to:

    a. Collect data through careful observations.
    b. Search for patterns of events (classify).
    c. Infer cause-effect relationships.
    d. Construct a statement about what you think a future observation will be, based on the pattern of events (predict).
    e. Test the dependability of the prediction.

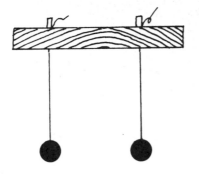

This next activity involves the use of pendulums. (With pendulums your students can have fun and learn important concepts at the same time! See the Delta Education unit, *Pendulums* for many activities provided for investigation and keeping interest high.)

    ⇨ **Go** to the prediction supply area, pick up a pendulums support, and set up your pendulum like those shown.

Set one of the pendulums in motion and begin exploring *round -trips*. A pendulum makes a *round-trip* when it is set in motion and returns to its starting position.

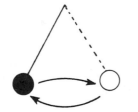

In your explorations, try to find out what affects the length of time it takes to make a round-trip. Check the things in this list that you feel affect the round-trip time:

1. _____ the size of the bob

2. _____ the mass of the bob

3. _____ the distance of the swing

4. _____ the length of the pendulum

5. _____ the type of path (back and forth, circle, and so on.)

Compare your answers with someone else's or check your answers with the following.

SELF-CHECK ✔

You probably found that the round-trip time of a pendulum is most affected by the *length* of the pendulum.

*A*ctivity 4

MAKING DEPENDABLE PREDICTIONS

While you were exploring with pendulums, you were making observations and gathering data about pendulums and motion. Through exploration and observation you began making inferences as to what might affect round-trip time for the pendulum. In the next activity you will concentrate on one feature, the length of the pendulum, and begin searching for patterns. Discovering *patterns* will enable you to make dependable predictions about the behavior of pendulums. By improving your inferences about what affects a pendulum's swing you increase the likelihood that your predictions are correct. In other words, you are building *confidence* in your predictions.

You may also increase the amount of confidence you have in a prediction by arriving at the same predicted values by different methods. The closer the agreement between predicted values arrived at by different methods, the greater the confidence you may have in the prediction.

In this next activity you will observe what happens when the length of a pendulum is systematically changed. Your observations will then be a basis for making predictions about the motion of the pendulum. Follow directions carefully.

Step 1: In this part of the activity you will be working with columns 1 and 2 of the chart that follows. Adjust the length of the pendulum to 15 cm (measure to the middle of the bob). Count the number of full swings in a 30 second interval. Record the number of full swings in the blank space in column 2.

Repeat the procedure for 25 cm, 35 cm, and 45 cm. These are the observed number of full swings for those pendulum lengths.

How Does Varying the Length of a Pendulum Affect Its Motion?

(1) Length (cm)	(2) Observed Swings	(3) Predicted by Method 1	(4) Predicted by Method 2	(5) Observed Swings
15				
20				
25				
30				
35				
40				
45				

Step 2: Examine columns 1 and 2. *Without swinging the pendulum*, predict the number of swings the pendulum would make in a 30 second interval for 20 cm, 30 cm, and 40 cm. Enter these predictions in the blank spaces in column 3. These represent your predictions using methods 1 (as used in Activity 2).

There is more than one way to make predictions using this data. This brings us to the next method —graphing.

Step 3: Using the data from columns 1 and 2, plot the data from columns 1 and 2 for 15 cm, 25 cm, 35 cm, and 45 cm, and on the following graph, draw a smooth curve through the plotted points.

How Does Varying the Length of a Pendulum Affect Its Motion?

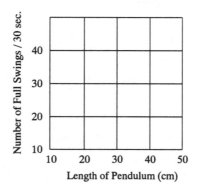

Compare your graph with the one below.

SELF-CHECK ✓

Your graph probably looks something like this:

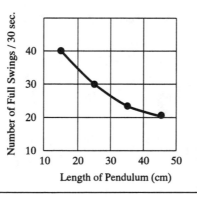

Step 4: By reading the graph, predict the number of swings for 20 cm, 30 cm, 40 cm.

Enter these predictions in the table under Column 4. These represent your predictions by Method 2—Graphing. (Predictions made between observed data are called *interpolations*. Predictions made beyond observed data are called *extrapolations*.)

Compare the values you predicted by Method 1 with those predicted by Method 2. The closer the agreement between the predicted values obtained by different methods, the greater the confidence you should have in your prediction. How confident are you that your predicted values are correct?

Step 5: Put the predictions to the test by observing the number of full swings of the pendulum at 20 cm, 30 cm, and 40 cm. Enter these numbers in the table under column 5. These are your observed values. Compare these to your predicted values. They should be fairly close.

SELF-CHECK ✓

Your table may look something like this:

———— How Does Varying the Length of a Pendulum Affect Its Motion? ————

(1) Length (cm)	(2) Observed Swings	(3) Predicted by Method 1	(4) Predicted by Method 2	(5) Observed Swings
15	40			
20		35	35	34
25	30			
30		27	27	27
35	25			
40		23	23	23
45	21			

Activity ♪

PREDICTION AND CHANCE

Few things are as dependable as *sunrise time* and the swing of a pendulum . . . not even the weather. Many factors can affect the accuracy of a prediction and often that factor is chance.

To illustrate this point, you will need the container of buttons. **Your instructor may have substituted beads or some other colored objects.** First, count the buttons to make sure that there are 25 red, 10 blue, 10 green, and 5 white ones. Because all the buttons are alike except for color, each button has the same chance to be taken as any other button (if taken without looking). Suppose you were to take a large sample (half of all the buttons) from the container, how many of each color would you predict?

Red   = _____

Blue = _____

Green = _____

White = _____

SELF-CHECK

Red   = 12 or 13
Blue   = 5
Green = 5
White = 2 or 3

The reason for the 12 or 13 and the 2 or 3 is that there are no half buttons in the container.

Why not try it? Take half the buttons from the container and see how they compare with your predictions.

Red   = _____

Blue   = _____

Green = _____

White = _____

SELF-CHECK

Compare your answers with someone else's. While they will vary, they probably are fairly close.

Often it is impossible to take a very large sample like half of the entire population because the population is very large and perhaps scattered over a very large area. Think how difficult it would be to observe the numbers of all the different kinds of plants in Yellowstone Park. What biologists do in situations such as this is to take many small samples.

Suppose you were to take a small sample of ten buttons from the container. Ten buttons are one fifth of all the buttons. How many red, blue, green, and white buttons would you predict to be in the sample? Enter the numbers in the table below beside *Predicted Results*.

	Red	**Blue**	**Green**	**White**
Predicted Results				
Your Sample				

Now take ten buttons from the container and enter the numbers of red, blue, green, and white buttons beside **Your Sample**. Compare your answers with the Self-check or someone else's answers.

SELF-CHECK ✓

	Red	Blue	Green	White
Predicted Results	5	2	2	1
Your Sample	3	4	2	1

Your answers will probably differ from our results and everyone else's.

If the predicted numbers were close to what you observed in your sample, you are probably satisfied with your accuracy and your *confidence* is high. If your prediction was not close to what you observed in your sample, you are probably tempted to try it again. The reason for the difference is that chance played a part in which buttons were selected in the sample.

Does the number of samples affect the accuracy of a prediction? Lets find out. In this activity, you will take five separate samples and see how closely the results compare with the predicted results.

1. Without looking, reach into the container and take a sample of ten buttons.
2. Record the number of red, blue, green, and white buttons under column labeled **Sample 1**.
3. Return the buttons to the container and mix them with the other buttons.
4. Repeat the procedure until you have completed all five samples.

Buttons	Sample 1	Sample 2	Sample 3	Sample 4	Sample 5	Total of Samples	Total Population
Red							25
Blue							10
Green							10
White							5

Each of the five individual samples probably differed from the predicted amounts as much as the first activity. Now total the number of red buttons taken in all five samples and record that number under total of samples for red. Repeat this process for the blue, green, and white samples.

How close were your total samples to predicting the total population of buttons? Compare your answer with ours in the Self-check.

SELF-CHECK ✓

Buttons	Sample 1	Sample 2	Sample 3	Sample 4	Sample 5	Total of Samples	Total Population
Red	6	4	6	6	3	25	25
Blue	4	3	2	1	5	15	10
Green	0	0	2	2	2	6	10
White	0	3	0	1	0	4	5

Were your answers closer, farther from, or about the same in predicting the correct amounts of red, green, blue, and white buttons?

Why did our answers differ? What do you suppose would happen to the accuracy of your prediction if you took ten samples instead of five?

Optional Activities for Predicting

ESTIMATING LARGE QUANTITIES

Have you ever wondered . . .

- How many blades of grass are in a lawn?
- How many words are in a book?
- How many grains of sand are in a beach?
- How many leaves are in a tree?
- How do they know how many eagles are alive?
- How many words are in a newspaper?
- How many stars are in the sky?
- How many grains of rice are in a bag?
- How many flakes are in a box of cereal?
- How many jelly beans are in that jar so you could win a bike?

As you can see, sometimes it is more practical to use estimates rather than actually counting. As a test for your ingenuity, devise and execute a method to determine the number of particles in a jar. The jar could be a liter of marbles, a jar of buttons, a jar of rice, or anything else your instructor is devious enough to contrive.

Compare your plans with those in the Self-Check, your instructor's or someone else's.

SELF-CHECK ☑

Listed below are some methods that may be used to predict large numbers of things. If you discover other methods, add them to the list.

1. Taking a Sample: Multiply the number of particles in a small sample by the number of sample-size quantities in the total.
2. Counting by Area and Column: Measure out a cupful of 250 mL; push the particles into a square; count the particles along one edge and square this number to get the number of particles the container held. Measure the number of containers held by the total sample and multiply this number of particles per container.
3. Weighing: Find the weight of a small, easy to count quantity and compare it to the weight of the whole.
4. Halving or Doubling: Divide the total quantity into two parts. Continue halving until you have a small enough quantity to count. Then double the number as many times as you halved the amount of particles.

Self-Assessment Predicting ✏️

I. Here is a graph showing the average monthly high temperatures for a particular city. Examine the graph and answer the questions that follow.

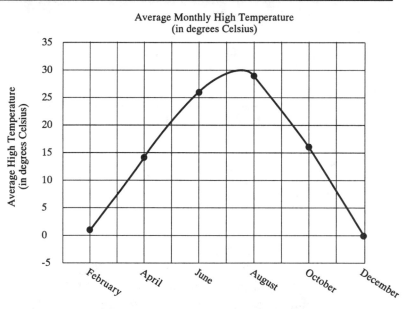

1. Through what basic process skill was the information in the graph first gathered? _____

2. Through what basic process skill was the information in the graph organized and put into order? _____

3. Through what basic process skill would you attempt to explain why the curve on the graph takes on this particular shape? _____

4. What was the average high temperature recorded for the month of February? _____ April? _____

5. Through what process would you forecast the average high temperature for the month of May? _____

6. Predict the average high temperature for the month of March: _____
   January: _____

7. Where in the United States would you predict this city might be located:

   _____

8. Write a statement comparing the *confidence* you have in your predictions in item #6 with the confidence you have in your prediction in item #7.

   _____

II.   1. Kim had a jar containing 100 pennies. Twenty-five of the pennies were dated 1992, fifty of the pennies were dated 1993, and the remaining pennies were dated 1994 Without looking, Kim took ten pennies from the jar and examined their dates. How many pennies would you predict that she found for each of the years?

       1992 = _____ pennies
       1993 = _____ pennies
       1994 = _____ pennies

   2. When Kim looked at the pennies, she found that five pennies were dated 1992, four pennies were dated 1993, and one penny was dated 1994. If Kim were to obtain a more accurate idea as to the distribution of pennies for 1992, 1993, and 1994, what two things could she do?

   Compare your answers with the self-assessment answers.

___

SELF-ASSESSMENT ANSWERS

I.   1. <u>Observation</u>
      2. <u>Classification</u>
      3. <u>Inference</u>
      4. <u>about 2°C, about 14°C</u>
      5 <u>Prediction</u>
      6. <u>about 8°C, about 1°C</u>
      7. The city was actually Detroit, Michigan
      8. You should have felt much more confident in the predictions you made in item #6 than the prediction you made for item #7. The difference lies in the fact that your predictions for #6 were based upon careful and comprehensive observations and a definite pattern appears in the data. Your prediction for item #7 was most likely a guess.

II.   1. Answers will vary somewhat, but with luck you could expect two or three dated 1992, five dated 1993, and two or three dated 1994.
      2. Kim could take larger samples (half of the pennies) or many small samples of the pennies.

IDEAS FOR YOUR CLASSROOM

1. *Food Webs.* Here is a simple food web.

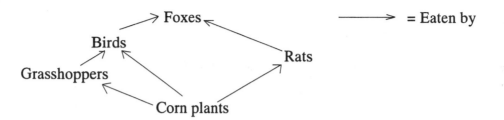

   Predict what would happen if any one organism were removed from the web.

2. *Electromagnetism.* Construct an electromagnet by wrapping about ten coils of electrical wire around a nail and connecting the ends of the wire to the opposite ends of a flashlight battery. Count the number of paper clips the electromagnet can pick up. Predict ways to increase the strength of the electromagnet.

3. *Inclined Planes.* Make an incline by placing a book under one end of a ruler. Place a marble in the groove of the ruler at the top of the incline. Release the marble and measure the distance the marble travels. Predict how the distance the marble travels can be increased.

4. *Electrical Conductors and Insulators.* Construct an electrical circuit by connecting the opposite ends of a flashlight battery to a flashlight bulb using two wires. Be sure the circuit works (the bulb lights). Test various objects to see if they are good conductors of electricity by first making a gap in the circuit between the bulb and one wire and inserting the object to be tested in the gap. If the bulb lights the object is a conductor. With a new set of objects made of various materials (glass, rubber, plastic, wood, different kinds of metals) predict which ones are conductors.

5. *Sound.* Add different amounts of water to several soda bottles and predict which will produce the highest pitch when struck sharply with a pencil. Which will give the highest pitch when someone blows into the top of the bottle?

6. *Combustion.* Invert jars of varying volumes over burning birthday candles. Predict which candle will burn the longest.

7. *Sound.* Hammer nails in pairs at various distances apart on a board. Stretch a rubber band over each pair of nails. Predict which rubber band will make the highest pitch when plucked. Predict a way to change the pattern.

THOUGHT-STARTER QUESTIONS

- What do you think will happen if . . . ?
- Predict what might happen next.
- If this is changed, what will happen to that?
- How will changing this variable affect that variable?
- What do you think might happen next?

SELECTED RESOURCES FOR TEACHING ELEMENTARY AND MIDDLE SCHOOL SCIENCE

Activities Integrating Mathematics and Science (AIMS)
Grades K–8
P.O. Box 8120
Fresno, California 93747, 209/255–4094

Bottle Biology
Grades K–6
Kendall/Hunt Publishing Company
4050 Westmark Drive
Dubuque, IA 52004, 800/258–5622

Chemical Education for Public Understanding (CEPUP)
Grades 6–8
Addison Wesley Publishing Company
2725 Sand Hill Road
Menlo Park, CA 94025, 800/447–2226

Elementary Science Study (ESS)
Grades K–8
Delta Education, Inc.
P.O. Box 915
Hudson, NH 03061 800/442–5444

Full Option Science System (FOSS)
Grades 3–6 (K–2 under development)
Encyclopaedia Britannica Educational Corporation
310 South Michigan Avenue
Chicago, IL 60604, 800/554–9862

Great Explorations in Math and Science (GEMS)
Grades K–10
Lawrence Hall of Science
University of California
Berkeley, CA 94720, 510/642–7771

GrowLab
Grades K–8
National Gardening Association
180 Flynn Avenue
Burlington, VT 05401, 802/863–1308

Science Activities for the Visually Impaired & Science Enrichment for Learners with Physical Handicaps (SAVI/ SELPH)
Grades 3–7
Center for Multisensory Learning
Lawrence Hall of Science
University of California
Berkeley, CA 94720, 415/642–8941

Science – A Process Approach II (SAPA)
Grades K–8
Delta Education, Inc.
P.O. Box 915
Hudson, NH 03061, 800/442–5444

Science for Life and Living:
Integrating Science, Technology and Health
Grades K–6
Kendall/Hunt Publishing Company
4050 Westmark Drive
Dubuque, IA 52004, 800/258–5622

Science and Technology for Children
Grades 1–6
Carolina Biological Supply Company
2700 York Road
Burlington, NC 27215, 800/334–5551

The Science Teacher, Science Scope, Science and Children
National Science Teachers Association (NSTA)
1840 Wilson Blvd.
Arlington, VA 22201-3000

Wisconsin Fast Plants
Grades K–12
Department of Plant Pathology
University of Wisconsin–Madison
1630 Linden Drive
Madison, WI 53706, 608/263–2634

ASSESSING FOR SUCCESS: PERFORMANCE TASK[1]

Preparation

Station Two - Water on Objects

Materials (per station)
- ✓ five 5-cm square pieces of white or tan paper napkins for each class
- ✓ five 5-cm square pieces of buff manila folder for each class
- ✓ five 5-cm square pieces of white or tan paper towel for each class
- ✓ one 5-cm square piece of white, unlined index card
- ✓ one clear plastic sealable sandwich bag
- ✓ one eye dropper
- ✓ one hand lens
- ✓ one small container (100 mL)
- ✓ one large container (150 - 250 mL)
- ✓ paper towels
- ✓ one non-water-soluble ink marker or stamp pad
- ✓ and a direction sheet for Station Two

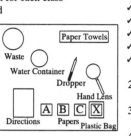

Preparation
1. Label each 5-cm square piece of white or tan paper napkin "A."
2. Label each 5-cm square piece of buff manila folder "B."
3. Label each 5-cm square piece of white or tan paper towel "C."
4. Label each piece of 5-cm square piece of white, unlined index card "X."
5. Place "X" in the plastic sandwich bag and seal it.
6. Label the bag "Do Not Open."
7. Label the 100-mL container "Water," and fill with 50 mL of water.
8. Label the large container "Waste."
9. Tape the direction sheet to the lower left side of the tabletop.
10. Place all materials on the tabletop as shown.
11. Make sure that fresh sets of papers A, B, and C are available for every student who will be tested at the station.

Directions to the Students

1. Check the materials

- ✓ Three pieces of paper (marked A, B, and C)
- ✓ Plastic bag containing a paper marked X
- ✓ Hand lens
- ✓ Dropper
- ✓ Container of water
- ✓ Container marked "Waste"

2. Read question 1 on the answer sheet for **Station Two.**
3. Use the pieces of paper marked A, B, and C only. **(Do not open the plastic bag.)**
4. Place **one** drop of water from the container on each piece of paper. Use the hand lens to look at what happened to each drop of water.
5. Answer question 1 on your answer sheet.
6. Look at the paper marked X inside the bag. **Do not open the bag!** You may look at paper X with the hand lens.
7. Answer questions 2 and 3 on your answer sheet.
8. Put the used pieces of paper marked A, B, and C in the container marked "waste."

Answer Sheet

1. What happened to the drop of water on each piece of paper?

   On paper A, the drop of water _____

   On paper B, the drop of water _____

   On paper C, the drop of water _____

2. You can not put water on Paper X. But **if** you could, **predict** what would happen to the drop of water. _____

3. Why did you predict this would happen? _____

Scoring Procedure

For question 1: Maximum score is 3 points
   Sample of Acceptable Answers:
   - On papers A and C, the drop of water was absorbed; soaked in; spread out; got bigger; expands; fills up/makes squares or blocks; goes through.
   - On paper B, the drop of water was not absorbed; sits on top; stays in ball; stays the same; doesn't spread: stays a drop; won't go through; bubbles on top.

For question 2: Because paper X is similar to paper B, the student prediction must indicate that he or she has observed the similarity of the two papers.
   Sample of Acceptable Answers:
   - the same thing that happened to paper B; it would not be absorbed; the water would sit on top.

For question 3: The students explain the prediction made for question 2, such as:
   - the paper is hard; paper X is shiny; it has no holes; looks like paper B

1. Doran, R.L., Reynolds, D., Camplin, J., and Hejaily, N. (1992) Evaluating Elementary Science. *Science and Children*, 30(3), 33-35, 63-64.

DECISION MAKING 1

Now that you have learned the Basic Science Process Skills, you can use what you have learned to improve existing science curricula. In learning the science process skills, you not only mastered the skills, but you also learned something about how these skills can be learned. By using this knowledge you can begin making some important instructional decisions about teaching science, especially the science process skills. In this section you will focus on the application of what you know about the science process skills to improve elementary school science textbook activities. The decisions you make can significantly enhance the quality of science in which your students are engaged.

Read *Textbook Activity Example A* on the next page. Think about how you might change the activity and the suggested teaching strategies to better emphasize the science process skills.

As you study the sample activity, look at both the content and skills your students will be learning and how they would be learning them.

Ask yourself, *How will I provide opportunities for my students to:*

- *use their senses?*
- *classify and form concepts?*
- *measure and quantify their descriptions of objects and events?*
- *communicate orally and in writing what they know and are able to do?*
- *infer explanations and change inferences as new information becomes available?*
- *predict possible outcomes before they actually observe?*

On a separate piece of paper, with these questions in mind, write what you consider to be strengths and weaknesses of *Example A*. Then consider how you might change the activity to improve the weak areas.

After you have studied *Example A* and thought about how you might change it, turn the page and look at the annotated version of *Example A*. The changes made to *Example A* are only a few modifications that could be made to this activity to better emphasize the process skills. Your ideas for modifying this activity may have been different and even better.

Textbook Activity Example A

ACTIVITY

Science Skills

observing, collecting data, making a graph

Fruits and Seeds

How many seeds are in some fruits?

1. Remove all the seeds from one fruit and place them on a paper towel.
2. Count and record the number of seeds.
3. Do steps 1 and 2 for each fruit.
4. Make a graph to show the number of seeds in each fruit.

Which fruit has the most seeds?
Which fruit has the least seeds?

TO THE TEACHER

Time: 30 min.

Groups: 4 students per group

Materials:
- ✓ 4 different fruits per group
- ✓ spoons
- ✓ paper towels
- ✓ graph paper

Objective:
Students will count the seeds in a variety of fruits and graph the results.

Lesson Setup:

- Use a variety of fruits, some with many seeds and some with just one.
- Cut the fruit for the students.
- Every student should make a graph.

Suggestions:

- Students should use the spoons to remove the seeds.
- Remind students to keep fruits and seeds on the paper towels.
- Show students examples of graphs
- Check to see that students are constructing and labeling graphs accurately.

Textbook Activity Example A

Here are some modifications to this activity that *we* made.
Your modifications may be even better.

ACTIVITY

Use 6 Fruits
- *Large Grapefruit*
- *Apple*
- *Cucumber*
- *Peach*
- *Orange*
- *Muskmellon*

Science Skills

Communicating, Classifying, observing, collecting data, making a graph, *predicting*

Fruits and Seeds

How many seeds are in some fruits?

1. Have the students name the fruits and classify them according to size.

2. Have students find other ways to classify them.

3. Have students predict how many seeds are in each & record their predictions on a piece of paper. Help them design a chart.

4 ~~1.~~ Remove all the seeds from one fruit and place them on a paper towel.

5 ~~2.~~ Count and record the number of seeds. *Compare with their predictions*

6 ~~3.~~ Do steps 1 and 2 for each fruit.

7 ~~4.~~ Make a graph to show the number of seeds in each fruit.

8. As a class, make a large graph compiling the data from each group

Which fruit has the most seeds?
Which fruit has the least seeds?

9. Ask each student to write what he or she found out using a word bank on the board

TO THE TEACHER

Time: ~~30~~ 40 min.

Groups: 4 students per group

Materials:
✓ ~~6~~ ~~4~~ different fruits per group
✓ spoons
✓ paper towels
✓ graph paper

Objective:
Students will count the seeds in a variety of fruits and graph the results.

Lesson Setup:

- Use a variety of fruits, some with many seeds and some with just one.
- Cut the fruit for the students.
- Every student should make a graph.

Suggestions:

Fruit | Number of seeds
Predict | Count

- Students should use the spoons to remove the seeds.
- Remind students to keep fruits and seeds on the paper towels.
- Show students examples of graphs
- Check to see that students are constructing and labeling graphs accurately.

Safety Tips: Remind students not to eat the fruit. Ask students to wash their hands when finished.

Here is another textbook activity example. Your task is to modify this activity to emphasize the process skills as modeled in the previous example. It may help you to review the questions on page 109 and to note the strengths and weaknesses of this activity. Then make your changes right on this activity page. When you are done, see the next page for modifications we made.

Textbook Activity Example B

ACTIVITY

In What Order Do the Parts of a Young Plant Develop?

You Will Need:
Seeds: corn and bean (vine beans work best), soaked in water overnight: paper towels, staples, water, plastic sandwich bags

Follow this Procedure:
Use your thumbnail to split open one seed and observe the inside.

Embryo Plant — Food for Plant — Seed Coat

Make a seed germinator by following these directions:
1. Fold a paper towel and slip it into the plastic sandwich bag to line the bag.
2. Make a row of staples about 4 cm from the bottom of the bag to form a shelf on which the seeds will sit.
3. Place about five seeds inside the bag just above the row of staples.
4. Holding the bag upright, slowly add water to moisten the paper towel and allow a little water to accumulate below the staples. It will be important to keep the towel moist all during germination.
5. Close the bag, and tack it to the bulletin board where it can be easily observed. If the surface on which you attach the bag is porous, you may want to cover it first with aluminum foil as moisture may accumulate on the back of the bag.

The seeds should begin to sprout within a few days. Record observations of the seed each day.

What are the parts of the growing young plant?

In what order do you see the parts develop?

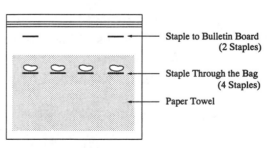

Staple to Bulletin Board (2 Staples)

Staple Through the Bag (4 Staples)

Paper Towel

Textbook Activity Example B

Here are some modifications to this activity that *we* made.
Your modifications may be even better.

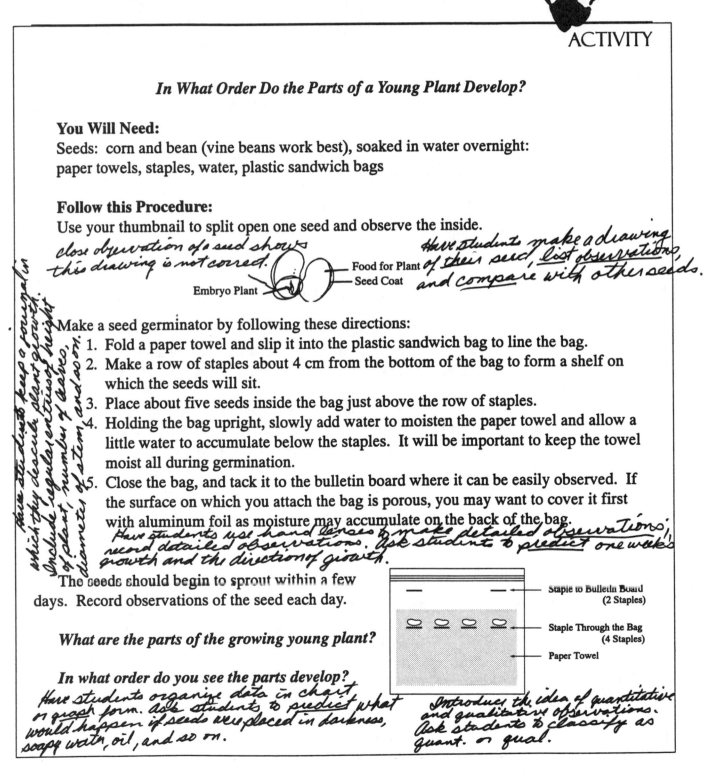

ACTIVITY

In What Order Do the Parts of a Young Plant Develop?

You Will Need:
Seeds: corn and bean (vine beans work best), soaked in water overnight:
paper towels, staples, water, plastic sandwich bags

Follow this Procedure:
Use your thumbnail to split open one seed and observe the inside.

close observation of a seed shows this drawing is not correct.

Have students make a drawing of their seed, list observations, and compare with other seeds.

— Food for Plant
— Seed Coat
Embryo Plant

Have students keep a journal in which they describe plant growth. Include together root, height of plant, number of leaves, diameter of stem and so on.

Make a seed germinator by following these directions:
1. Fold a paper towel and slip it into the plastic sandwich bag to line the bag.
2. Make a row of staples about 4 cm from the bottom of the bag to form a shelf on which the seeds will sit.
3. Place about five seeds inside the bag just above the row of staples.
4. Holding the bag upright, slowly add water to moisten the paper towel and allow a little water to accumulate below the staples. It will be important to keep the towel moist all during germination.
5. Close the bag, and tack it to the bulletin board where it can be easily observed. If the surface on which you attach the bag is porous, you may want to cover it first with aluminum foil as moisture may accumulate on the back of the bag.

Have students use hand lenses to make detailed observations; record detailed observations. Ask students to predict one week's growth and the direction of growth.

The seeds should begin to sprout within a few
days. Record observations of the seed each day.

— Staple to Bulletin Board
(2 Staples)
— Staple Through the Bag
(4 Staples)
— Paper Towel

What are the parts of the growing young plant?

In what order do you see the parts develop?

Have students organize data in chart or graph form. Ask students to predict what would happen if seeds were placed in darkness, soapy water, oil, and so on.

Introduce the idea of quantitative and qualitative observations. Ask students to classify as quant. or qual.

MODIFYING REAL TEXTBOOK ACTIVITIES

The textbook examples you have just studied are typical of the kinds of science activities you might find in an elementary school textbook. To gain a little more experience at modifying materials to emphasize the science process skills and to become acquainted with real textbook activities, you have one more task to complete.

Obtain an elementary science textbook for grade K, 1, 2, or 3. Locate an activity and modify it to better emphasize the basic science process skills just as you did in the previous examples. You might refer again to the questions on page 109 that focus on these skills. You may find it helpful to think of this task as a three-step procedure:

1. Examine the activity.
2. Identify parts that could be improved.
3. Improve it.

For feedback on your modifications, see your instructor, or try your modified activity with children and assess their skills. When you have completed this task, you will be ready to learn the Integrated Science Process Skills in Part Two of the book.

Integrated Science Process Skills

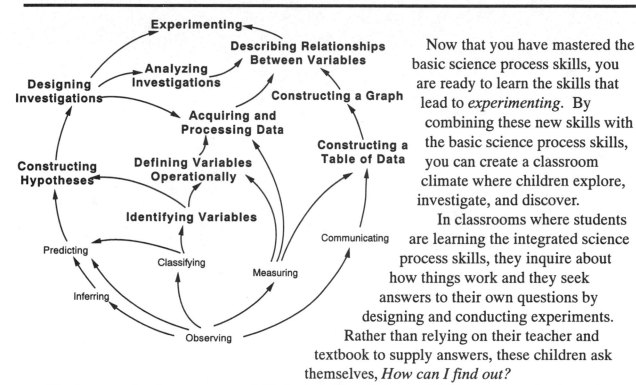

Now that you have mastered the basic science process skills, you are ready to learn the skills that lead to *experimenting*. By combining these new skills with the basic science process skills, you can create a classroom climate where children explore, investigate, and discover.

In classrooms where students are learning the integrated science process skills, they inquire about how things work and they seek answers to their own questions by designing and conducting experiments. Rather than relying on their teacher and textbook to supply answers, these children ask themselves, *How can I find out?*

The integrated science process skills include identifying variables, constructing hypotheses, analyzing investigations, tabulating and graphing data, defining variables, designing investigations, and experimenting. Learning these skills empowers students to answer many of their own questions. Students who have learned the integrated skills have the tools to interpret what they observe and to design investigations to test their ideas.

The integrated skills are not separate and distinct from the basic skills. The basic skills provide the foundation for the more complex integrated skills. For example, the predicting skills you learned in Part One are used to construct hypotheses. In Part Two you will learn that a hypothesis is a special kind of prediction that sets the stage for investigating relationships in science. You will find, as your students also will, that experimenting leads to asking more questions and conducting more experiments. Experimenting is a form of problem solving that requires the integration of all the other thinking skills as illustrated above.

Part Two of this book begins with the integrated science process skill of *Identifying Variables*. Each time you learn a new skill, ask yourself the same questions you asked in Part One:

How am I learning this skill?
How will I teach this skill to students?

Teaching Children

After you have completed the Integrated Science Process Skills, you will be asked to make some instructional decisions about how you might teach these same skills to your students.

117

Identifying Variables

CHAPTER 7

Sight

Smell

Sound

Taste

Touch

PURPOSE

In this chapter, you will be learning one of the skills needed when conducting an investigation. This important skill will be used throughout this section whenever you analyze how someone else conducted an investigation or whenever you plan and carry out an investigation of your own.

OBJECTIVE

After studying this chapter you should be able to:
1. identify the variables in a written statement or description of an investigation.
2. classify the variables as manipulated or responding.

Approximate time for completion: 40 minutes

MEASUREMENT SKILLS

Throughout the exercises in this section, you will be asked to answer questions that require you to make measurements. It is assumed that you know how to measure mass, length, temperature, and volume. If you are not sure that you know how to make these measurements, you may wish to review Chapter 4.

The best way to become comfortable with science is to do science. You need to investigate a bit, use some equipment, and get your hands dirty. To accomplish this and also to learn how to identify variables, carry out the following activities.

⇨ *Go* to the supply area and locate the following items:

✓ safety goggles
✓ 4 small identical containers (baby food jars or plastic cups)
✓ thermometer
✓ plastic spoon
✓ large container (about a liter for holding water)
✓ calcium chloride (chemical used to control ice on roads)
✓ graduated cylinder

Activity 1

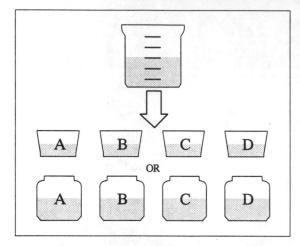

Fill a large container with tap water that is about room temperature. Use the graduated cylinder to fill each container with 75 mL of water.

Activity 2

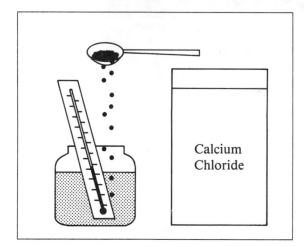

Measure the temperature of the water in one of the containers (call it container A). Add one level scoop of calcium chloride and stir it until it dissolves.

Measure the temperature of the water as soon as the calcium chloride dissolves.

1. What was the initial temperature of the water **before** adding the calcium chloride?

    _____

2. What was the temperature of the water **after** adding the calcium chloride?

    _____

3. What happened to the temperature of the water in the container?

    _____

4. How many degrees did the temperature change when you added one level scoop of calcium chloride? _____

You probably found that the temperature increased about 3 to 6 degrees Celsius. The temperature could be more or less than this, depending on the amount of water you used and the amount of calcium chloride.

To keep track of your measurements, record them in the table shown below. You should record both the number of scoops of calcium chloride added and the **change in temperature** for each container.

Container	Number of Scoops of Calcium Chloride	Temperature Change (°C)
A		
B		
C		
D		

Activity **3**

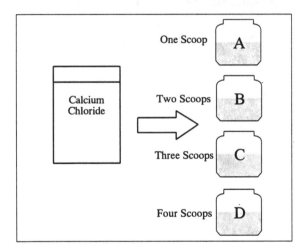

Continue the procedure given in Activity 2 for B. This time add two level scoops of calcium chloride. Again, determine how much the temperature changes from the initial temperature of the water. Record your data in the table.

Repeat the procedure for C and D using the amounts of calcium chloride shown in Activity 3.

Your table of data should be complete now. It should look similar to the one that follows except for the last column. We recorded the change in temperature that we noticed when we did the experiment. Your numbers will probably be different, depending on the amount of water in your container and the size of the scoop.

Our Data

Container	Number of Scoops of Calcium Chloride	Temperature Change (°C)
A	1	5
B	2	9
C	3	14
D	4	21

Answer the following questions about what you did:

5. Did you use the same amount of water in each container? _____

6. Did you use the same amount of calcium chloride in each container? _____

7. Did the temperature change the same amount for each container? _____

8. What prediction would you make if you added five scoops of calcium chloride to the same amount of water? _____

If you followed the directions carefully, you should have used the same amount of water in all four containers. You should have added different amounts of calcium chloride to each container (1 scoop to container A, 2 to container B, 3 to container C, and 4 to container D.) The temperature should have increased by different amounts in the containers. The increase was greater for those containers in which more calcium chloride was dissolved. Finally, you probably predicted that the temperature change would be even greater if five scoops of calcium chloride were added (we predicted a 27 degree change.)

In the last part of this chapter, you will complete a short unit of programmed instruction. Perhaps you have never used this type of instruction before. The procedure is very simple. The instruction is broken up into small pieces. These pieces are presented as frames. Information is given in each frame and you, the learner, are expected to give a response. After writing your response, you can check its accuracy by comparing your response with the answer that follows in each frame. Use a sheet of paper to cover the answers and slide the paper down the page as you read. After each question, a special symbol ✓ will signal you that the answer to the question will appear next. Keep the answer covered until you have responded.

The purpose of this programmed instruction is to teach you to identify **variables**. After you have learned to identify variables, you will learn to classify them as **manipulated** or **responding** variables. Manipulated and responding variables are also called **independent** and **dependent** variables.

Read this statement:

The height of bean plants depends on the amount of water they receive.

In this statement two variables are described:

1. height of bean plants
2. amount of water

Read the following frames and record your responses.

A **variable** is something that can vary or change. What are the variables in this statement?

The time it takes to run a kilometer depends on the amount of exercise a person gets.

1. _____   _____

_____

      time to run a kilometer   amount of exercise   ✓

It would not be correct to name just *time* or *exercise* as the variables. You must include how each variable will be measured or described. For example, height of plant, number of fruit produced, color of leaves, and diameter of the stem are all variables.

What are the variables in this statement?

The higher the temperature of water, the faster an egg will cook.

2. _____   _____

_____

      temperature of water   time needed for an egg to cook   ✓

Your answers don't have to be exactly the same as these but they should be close.

What are the variables in this investigation?

An investigation was done to see if keeping the lights on for different amounts of time each day affected the number of eggs chickens laid.

3. _____   _____

_____

      hours (or amount) of light   number of eggs   ✓

Again, your answers do not have to be exactly like these, but they should be similar. Just *light* and *eggs* however, would be incorrect descriptions of the variables.

What are the variables in this statement? Remember a variable is something that can change or vary.

The temperature of the water was measured at different depths of a lake.

4. _____   _____ ✓

        temperature of water   distance below the surface

Perhaps you said, *depth of lake* for the second variable.

Here is another statement. Identify the variables in it.

Grass will grow taller if it is watered a great deal and if it is fertilized.

5. _____   _____   _____ ✓

    height of grass   amount of water   amount of fertilizer

Many variables can be included in a statement. You may find one, two, three, or more.

Think back to the investigation you did at the beginning of this chapter. Then complete this statement:

If the amount of calcium chloride added to the water increases, the temperature of the water (increases, decreases)

6. _____

                increases ✓

Clearly, the more scoops of calcium chloride, the more the water temperature increased.

What were the variables in the investigation on calcium chloride you carried out?

(HINT: look back in the statement in Frame 6)

7. _____   _____
   number of scoops of calcium chloride   temperature change of water ✓

You might have said *amount of calcium chloride* or *temperature of water*. These answers would be correct also.

If a variable is deliberately changed, it is called a **manipulated variable**. Which of the two variables was the manipulated variable in your investigation?

8. _____
   number of scoops of calcium chloride ✓

You deliberately used a different number of scoops for each container so you manipulated this variable.

What variable was manipulated in this investigation?

The amount of pollution produced by cars was measured for cars using gasoline containing different amounts of lead.

9. _____
   amount of lead in gasoline ✓

What is the manipulated variable in this statement?

Lemon trees receiving the most water produce the largest lemons.

10. _____
   amount of water ✓

The *amount of water* could be manipulated or changed to determine its effect on the size of the lemons produced.

Identify the manipulated variable in the following:

The amount of algae growth in lakes seems to be directly related to the number of bags of phosphate fertilizer sold by the local merchants.

11. _____

_____

number of bags of phosphate fertilizer sold ✓

An investigation was performed to see if corn seeds would sprout at different times depending on the temperature of the water in which they were placed.

12. _____

_____

temperature of water ✓

In each of the last four frames, one variable was changed to see what would happen. A variable that is deliberately changed is called a **manipulated variable**.

The more water you put on grass, the taller it will grow.

Amount of water is the manipulated variable in the above statement. The other variable is the *height of grass*. The variable that may change as a result of changing the manipulated variable is called the **responding variable**.

Identify the **manipulated** and **responding** variables in this statement:

More bushels of potatoes will be produced if the soil is fertilized more.

13. Manipulated variable: _____

Responding variable: _____

_____

amount of fertilizer (manipulated variable) ✓

number of bushels of potatoes (responding variables)

The amount of fertilizer could be *manipulated* to see if the number of bushels of potatoes *responded*.

Think back to the investigation you did at the beginning of this chapter. You manipulated the amount of calcium chloride. What was the responding variable?

14. _____

_____ ✓

temperature change of water

In each container, you manipulated the amount of calcium chloride to see if the temperature of the water would respond.

Look at the sketch. It shows an investigation similar to the one you did. Notice that different amounts of water are used in each container, with one scoop of calcium chloride added to each container. After the calcium chloride dissolves, the temperature change in each container will be determined.

PUT *ONE* SCOOP IN EACH CONTAINER

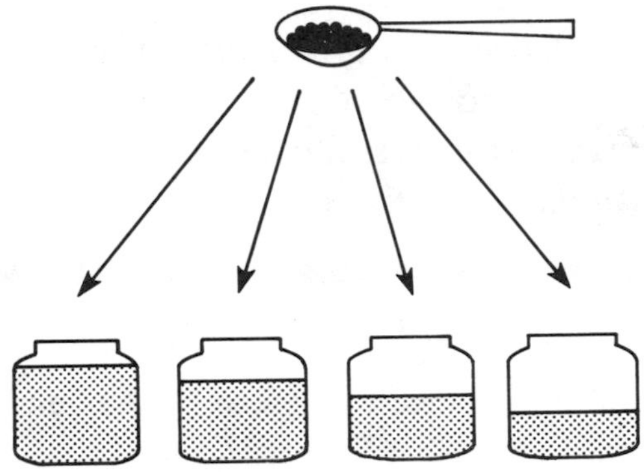

What are the manipulated (MV) and responding (RV) variables in this investigation?

15. MV _____

RV _____

_____ ✓

(MV) amount of water
(RV) temperature change of water

This is similar to the investigation you did, but now a different variable is being manipulated.

What are the manipulated and responding variables in this investigation?

Five groups of rats are fed identical diets except for the amount of Vitamin A that they receive. Each group gets a different amount. After three weeks on the diet, the rats are weighed to see if the amount of Vitamin A received has affected their weight.

16. MV _____

    RV _____

(MV) <u>amount of Vitamin A</u>
(RV) <u>weight of rats</u>

If the *amount of Vitamin A* is manipulated or changed, then perhaps the *weight of rats* will respond. Of course, weight may not be affected if Vitamin A is not essential. The *weight of rats* is still the responding variable whether or not it is actually affected by the manipulated variable.

An experiment was done with six groups of children to see if scores on their weekly spelling tests were affected by the number of minutes of spelling practice they had each day.

17. MV _____

    RV _____

(MV) <u>minutes of spelling practice</u>, (RV) <u>score on spelling test</u>

Will the number of nails picked up by an electromagnet be increased if more batteries are put in the circuit?

Suppose an investigation was carried out on the problem above. What would the variables be?

18. MV _____

    RV _____

(MV) <u>number of batteries in circuit</u>, (RV) <u>number of nails picked up</u>

In this chapter, you have learned to identify variables. You have also learned that a *variable* is something that can change or vary and that there are two different kinds of variables. A variable that is deliberately changed is called a *manipulated* variable. The variable that may change as a result of changing the manipulated variable is called the *responding* variable.

Now take the Self-assessment for Chapter 7 and check your answers.

Continue on to Chapter 8 or return to Chapter 7 for more study based on the results of your Self-assessment.

Self-Assessment Identifying Variables

IDENTIFYING VARIABLES

For each of the following statements or descriptions identify the manipulated variable (MV) and responding variable (RV). The answers will follow.

1. Students in a science class carried out an investigation in which a flashlight was pointed at a screen. They wished to find out if the distance from the light to the screen had any effect on the size of the illuminated area.

   MV _____

   RV _____

2. The number of pigs in a litter is determined by the weight of the mother pig.

   MV _____

   RV _____

3. The State Agriculture Department has been counting the number of foxes in Brown County. Will the number of foxes have any effect on the rabbit population?

   MV _____

   RV _____

4. The score on the final test depends on the number of subordinate skills attained.

MV _____

RV _____

5. A study was done with white rats to see if the number of offspring born dead was affected by the number of minutes of exposures to X-rays by the mother rats.

MV _____

RV _____

SELF-ASSESSMENT ANSWERS

1. MV distance from light to screen
   RV size of illuminated area
2. MV weight of mother pig
   RV number of pigs in litter
3. MV number of foxes
   RV number of rabbits
4. MV number of subordinate skills attained
   RV score on final test
5. MV minutes of exposures to X-rays
   RV number of offspring born dead

ASSESSING FOR SUCCESS: STUDENT PORTFOLIO

Identifying variables is the first of several integrated science process skills students will learn to be able to eventually design and conduct their own experiments. Introducing student portfolios at this early point in instruction can evidence students' growth over time as they develop the necessary skills.

Two kinds of portfolios can be proposed to students - working folders and showcase folders. Working portfolios are holding bins for many pieces of work that is representative of learning still in progress. A showcase portfolio, on the other hand, evidences a student's best work. By periodically transferring pieces of a student's work from the working folder, the showcase portfolio gives the student, the parents, and the teacher a means to determine student performance levels based on his or her best work.

A showcase portfolio should include:

- a table of contents
- a student letter to the reviewer describing his or her portfolio
- a variety of entries that show what a student knows and is able to do.

Contents of a portfolio can include pretests, quizzes, tests, tables, graphs, laboratory reports, teacher notes, peer evaluations, and group designs of experiments. Because a portfolio is a collection of evidence, a science portfolio on the skills of experimenting should ultimately contain evidence of a student designed experiment. A written report of an experiment conducted at school or at home, or photographs of a science fair display of a science project can show mastery of the ability to design and analyze an experiment.

Portfolios encourage collaboration between teacher and students as they choose tasks and assignments that evidence their progress and achievement. Student self-evaluation is also fostered because students learn to monitor their own work and to reflect on their performance. The following are two examples of how self-evaluation can be encouraged.

**Science Research Checklist:
A Guide for Your Portfolio**

Provide evidence that you can do the following:

Planning
_____ I can choose my own topic.
_____ I can use several resources to find information on my topic.
_____ I can design or select an experiment based on my topic.

Conducting
_____ I can follow a procedure and use materials and equipment to conduct my experience.
_____ I can collect data from my experiment.

Communicating
_____ I can display my data as tables and graphs.
_____ I can explain what I found.
_____ I can write a report describing my experiment.

Portfolio Evaluation

Name _____
Unit/Activity _____
Date _____

Why I chose this piece of work:

_____
_____

What I learned:

_____
_____

What I want to learn next:

_____
_____

© Rezba, Sprague, Fiel, Funk, Okey, Jaus. **Learning and Assessing Science Process Skills**. Kendall/Hunt 1995.

Constructing a Table of Data

CHAPTER 8

PURPOSE

One of the skills needed to conduct an investigation is the organization of data in tables. When data are presented in well-organized tables, trends and patterns of change in data are often revealed.

OBJECTIVES

After studying the information in this chapter, you should be able to:

1. Construct a table of data when given a written description of the measurements made during an investigation.
2. Write data pairs from a table of data.
3. Match data pairs with points on a graph

Approximate time for completion: 35 minutes

For the programmed activities that follow remember to keep the answers covered until you have responded. The eye is sometimes quicker than the conscience.

When an investigation is conducted, the measurements made are called data. Measurements of time, temperature, and volume are examples of data. Organizing data into tables helps to see patterns in the results.

Although there are no absolute rules for constructing tables of data, there are conventions, or commonly agreed upon patterns of organization, that facilitate communication between the writer and the reader. For example, when constructing a table of data, the manipulated variable is recorded in the left column and the responding variable is recorded in the right column.

Column for the manipulated variable	Column for the responding variable

➤ 1. Read the following description of an experiment and identify the manipulated and responding variables. Record these variables in the table.

An investigation was conducted to see what happens to the height measured in centimeters of Wisconsin Fast Plants when 0, 1, and 2 of the plants' cotyledons were removed. (A cotyledon is the fleshy part of a seed, sometimes called a seed leaf.)

1. SELF-CHECK ✓

Number of Cotyledons Removed	Height of Plants (cm)

Notice that whenever units are used, they are included in the column heading as well.

When recording data in a table, the levels of the manipulated variables are ordered. Although data are sometimes ordered from largest to smallest, the usual procedure is to order data from smallest to largest. This organization establishes a pattern of change in the manipulated variables. If there is a corresponding pattern of change in the responding variable, it will be easier to recognize than if the levels of the manipulated variable were placed randomly in the table.

> 2. Using these number pairs (5,7) (3,3) (9,5) (1,6) (2,4) which of the following tables have been properly ordered?

A.		B.		C.		D.		E.	
MV	RV	MV	RV	MV	RV	MV	RV	MV	RV
5	7	3	3	1	3	1	6	9	5
3	3	2	4	2	4	2	4	5	7
9	5	9	5	3	5	3	3	3	3
1	6	1	6	5	6	5	7	2	4
2	4	5	7	9	7	9	5	1	6

2. SELF-CHECK

Only **D** and **E** are correct. In Table A the data are in random order and therefore incorrect. In B the data for the responding variable have been ordered. This is not correct, unless it occurs naturally as a result of ordering the data of the manipulated variable. In C the data pairs were separated, which is incorrect, and then ordered. Table E is correct, but the usual procedure is to order the data from smallest to largest as in Table D.

> 3. Try ordering the following data pairs in the table. Assume the first number in each pair is for the manipulated variable and the second number is for the responding variable.

(20, 17) (5, 18) (9, 12) (23, 26) (17,3) (27,32)

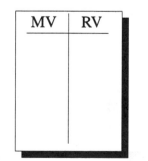

MV	RV

3. SELF-CHECK

Either is correct, but data for the manipulated variables are usually ordered from the smallest to the largest.

MV	RV
5	18
9	12
17	3
20	17
23	26
27	32

MV	RV
27	32
23	26
20	17
17	3
9	12
5	18

➤ 4. Try one more. Remember (MV, RV): (12,12) (13,10) (5,10) (8,12)

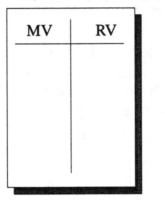

MV	RV

4. SELF-CHECK

Again, either is correct, but the first is preferred, where the data are ordered from smallest to largest.

MV	RV
5	10
8	12
12	12
13	10

MV	RV
13	10
12	12
8	12
5	10

➤ 5. Here is a written description of an investigation. Notice that some of the data have already been recorded in a table. Read the paragraph and record the remaining data.

The heights that balls bounced when dropped different distances were measured. A ball dropped 50 cm bounced 40 cm high. A 10 cm drop bounced 8 cm. A ball bounced 24 cm when dropped 30 cm. The bounce was 56 cm high for a 70 cm drop. A 100 cm drop bounced 80 cm.

Length of Drop (cm)	Height of Bounce (cm)
10	8

5. SELF-CHECK ✓

Length of Drop (cm)	Height of Bounce (cm)
10	8
30	24
50	40
70	56
100	80

When constructing tables of data, the manipulated variable [Length of drop (cm)] heads a column on the left, with all the levels listed below, usually from smallest to largest. The responding variable is recorded in the column to the right. Units of measurements are included in the column heading.

> 6. Here is another practice problem. Label the columns and record the data in the table provided.

The distance covered by a runner during each second of a race was measured. During the 15th second of the race the runner covered two meters. Three meters were covered during the 12th second. Four meters were covered during the 9th second. During the 6th second three meters were covered. During the 3rd second, two meters were traveled.

Time During Race (sec)	Distance Covered (m)

6. SELF-CHECK

Time During Race (sec)	Distance Covered (m)
3	2
6	3
9	4
12	3
15	2

MODIFYING A TABLE OF DATA TO RECORD REPEATED TRIALS

Most experiments should be repeated by testing each level of the manipulated variable several times. Repeated trials *increase confidence* in results by reducing the effects of chance errors that may occur in a single trial.

When repeated trials are conducted, the column for the responding variable is divided into smaller columns so data can be recorded for each repeated trial. Information, such as the average result or the range (how spread out the data are), is recorded in one or more columns to the right of the column for the responding variable. Information that is computed from data is called a *derived quantity*.

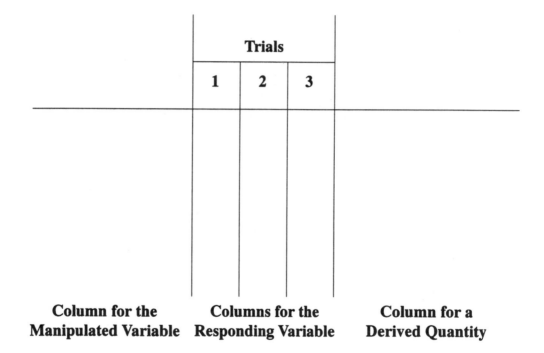

| Column for the Manipulated Variable | Columns for the Responding Variable | Column for a Derived Quantity |

➤ 7. Read the following description of an investigation and construct a table of data that has:

- columns for the manipulated variable, responding variable, and a derived quantity
- labels and appropriate units for each column
- subdivided column for the responding variable to reflect the number of repeated trials conducted.

An investigation was conducted to see how the number of minutes of heating time affected the temperature of water. A pan containing 1 liter of water was heated for 5 minutes. At the end of each minute the temperature in degrees Celsius was recorded. The experiment was conducted a total of 4 times.

7. SELF-CHECK ✓

Heating Time (min)	Temperature (°C) Trials				Average Temperature (°C)
	1	2	3	4	

WRITING NUMBER PAIRS FROM A TABLE OF DATA

Because graphs communicate data in pictorial form, they can show trends and patterns in data more effectively than a data table alone. Each point on a graph represents a pair of data. When writing data pairs, the value for the horizontal or X axis on a graph is written first, followed by the value for a graph's vertical or Y axis. The two numbers are separated by a comma and are placed in parentheses, for example (10,18). By convention, the manipulated variable is graphed on the horizontal (X) axis and the responding variable on the vertical (Y) axis.

If the convention for the placement of the MV and RV in data tables is followed, each data pair for a graph (MV, RV) can easily be determined from a table.

➤   8. Write the data pairs for constructing a graph from the following table of data.

Amount of Rain (cm)	Weight of Fruit (kg)			
45	125	(	,	)
55	140	(	,	)
60	150	(	,	)
66	200	(	,	)
70	280	(	,	)
75	310	(	,	)

8. SELF-CHECK ✓

(45,125)
(55,140)
(60,150)
(66,200)
(70,280)
(75,310)

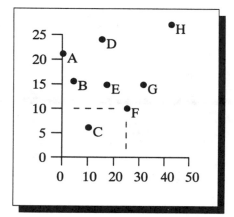

A data point on a graph is described using the values given on the horizontal and vertical axes. Imagine a **vertical** line drawn to the point labeled "F" on the graph. Such a line would intersect the X axis at the point with a value of 25 (halfway between the intervals of 20 and 30). Imagine a second line drawn **horizontally** to point "F". This line would intersect the Y axis at point with a value 10.

➤ 9. Match the letters of the remaining data points with the following number pairs:

(0, 21) ——————   (18, 15) ——————

(5, 16) ——————   (25, 10) ——————

(10, 6) ——————   (31, 15) ——————

9. SELF-CHECK ✓

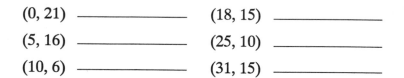

(0, 21) ___A___   (18, 15) ___E___

(5, 16) ___B___   (25, 10) ___F___

(10, 6) ___C___   (31, 15) ___G___

Being able to write data pairs from a table of data is the first step in being able to construct a graph. In the next chapter you will learn to make a graph from a data table.

Now take the Self-assessment for Chapter 8.

Self-Assessment Constructing a Table of Data

➤ 1. Construct an ordered table of data for the following. The length of shadows made by sticks of different length were measured. A stick 50 cm long made a shadow 40 cm long. The shadow was 5 cm long for a stick 10 cm. A 30 cm stick made a shadow 22 cm long and a stick 40 cm long made a shadow of 29 cm. The shadow was 12 cm for a stick 20 cm long.

Length of Stick (cm)	Length of Shadow (cm)

➤ 2. Construct a table to record data for this investigation. Include smaller columns for repeated trials as appropriate.

A study was conducted to see if amount of salt affected how fast the salt dissolved in water. Using the same amount and temperature of water, four different amounts of salt (10, 20, 30, and 40 mL) were stirred until no crystals were visible. Each amount of salt was tested three times.

➤ 3. Match the letter with the data pair that describes the location of each point.

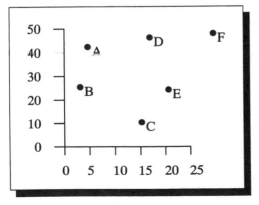

Data Pairs	Letter
(3, 26)	_____
(5, 42)	_____
(15, 10)	_____
(17, 45)	_____
(20, 25)	_____
(30, 50)	_____

SELF-ASSESSMENT ANSWERS

➤ 1.

Length of stick (cm)	Length of shadow (cm)
10	5
20	12
30	22
40	29
50	40

➤ 2. Amount of salt (mL)

Amount of Salt (mL)	Time to Dissolve (min) Trials			Average Time to Dissolve (min)
	1	2	3	
10				
20				
30				
40				

➤ 3.

Number Pairs	Letter
(3,26)	**B**
(5,42)	**A**
(15,10)	**C**
(17,45)	**D**
(20,25)	**E**
(30,50)	**F**

Student Assessment *Example* ☆

ASSESSING FOR SUCCESS: STUDENT/PEER/FAMILY CHECKLIST.

Name: _____ Date: _____

Data Table Title:

CRITERIA	SELF		PEER		FAMILY	
TABULATING DATA SKILLS	YES	NO	YES	NO	YES	NO
Does the title tell about the manipulated variable (MV) and the responding variable (RV)?						
Is the left hand column for the MV?						
Are the label and units given for the MV?						
Are the levels of the MV ordered?						
Is the right hand column for the RV?						
Are the label and units given for the RV?						
Is the RV column subdivided for repeated trials?						
Are the RV data correctly recorded?						
Are there additional columns for derived quantities such as the average and the range?						
Are the label and units given for the derived quantities?						
Are the derived quantities correctly calculated?						

© Rezba, Sprague, Fiel, Funk, Okey, Jaus, **Learning and Assessing Science Process Skills**. Kendall/Hunt 1995.

Constructing a Graph

CHAPTER 9

Sight

Smell

Sound

Taste

Touch

PURPOSE

A picture is worth a thousand words. Almost everyone has heard this famous saying. Often it is true that information can be communicated more easily with a picture instead of using a spoken or written message. Your task in this chapter is to learn to draw one special kind of picture - a graph.

OBJECTIVE

After studying this chapter you should be able to construct a graph when provided with a brief description of an investigation and a table of data.

Approximate time for completion: 45 minutes

An example of what you will be able to do when you finish this chapter is shown below. You will be given the type of information found on the left hand side of the page and will be expected to produce the type of material found on the right hand side.

You will be given:

INVESTIGATION: Beans were soaked in water for different lengths of time and their gain in mass was recorded.

Soaking Time (min)	Average Gain in Mass (g)
5	10
10	20
15	40
20	45
25	50
30	55

You will produce:

The Effects of Soaking Time On Mass of Beans

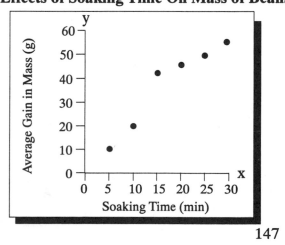

147

The part of the graph on the previous page, labeled *Average Gain in Mass*, is called the vertical or **y** axis and that part labeled *Soaking Time* is called the horizontal or **x** axis. Together they make up the axes of the graph.

In order to construct a graph from a table of data you must learn three skills:

1. Label the x axis with the manipulated variable and the y axis with the responding variable.
2. Determine an interval scale for each axis that is appropriate for the data to be graphed.
3. Plot the data pairs as data points on a graph.

Skill 1 **Label the X and Y Axes**

Amount of Fertilizer (kg)	Average Height of Plants (cm)
2	24
4	50
6	74
8	38

Suppose you wanted to construct a graph of the data given above. What would you have to do? After drawing the horizontal and vertical axes, write labels for the variables along these axes. When deciding which variable to assign to an axis, follow this rule:

The manipulated variable is always written along the horizontal axis. The responding variable is always written along the vertical axis.

In the case shown above, the amount of fertilizer was purposely manipulated and the height of the plants was then measured. So, *Amount of Fertilizer (kg)* should be written along the horizontal axis and *Height of Plants (cm)* should be written beside the vertical axis. The correct form is shown on the graph below:

The Effect of Amount of Fertilizer on Height of Plants

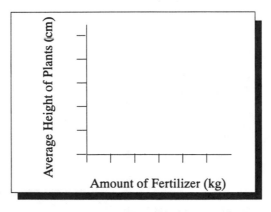

Descriptions of several investigations follow. Next to each description is a graph with the variables assigned to the axes. Your task is to determine whether the labels are correct.

How Does the Distance From Which a Ball Is Dropped Affect How High It Bounces?

➤ 1. *INVESTIGATION:* A ball is dropped from several distances above the floor and the height it bounces up is then measured.

    Labels OK ☐
    Labels reversed ☐

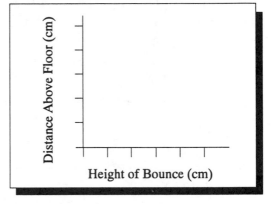

1. SELF-CHECK ✔

Labels are reversed.

Because the ball was deliberately dropped from different heights, that variable is manipulated and should be used to label the horizontal axis.

How Does the Volume of a Jar Affect the Burning Time of a Candle?

➤ 2. *INVESTIGATION:* A candle was burned under glass jars of different volumes to see if the length of time the candle burns is affected by the volume of the jar.

    Labels OK ☐
    Labels reversed ☐

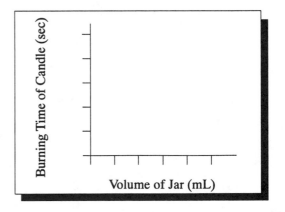

2. SELF-CHECK ✔

Labels OK

**How Does the Diameter of
a Siphon Affect Siphoning Time?**

➢ 3. *INVESTIGATION:* An investigation is done to see if the diameter of rubber tubing affects the time it takes to siphon water out of a container.

Labels OK ☐
Labels reversed ☐

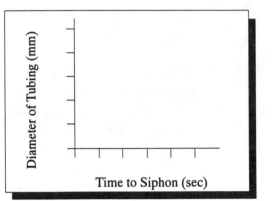

3. SELF-CHECK ✓

Labels reversed

Several descriptions of investigations follow. Write the variable being manipulated along the horizontal axis for each graph. Write the responding variable along the vertical axis. Be sure to indicate the appropriate measurement units for each variable.

**How Does the Gauge of Fishing Line Affect
the Number of Fish Caught?**

➢ 4. *INVESTIGATION:* A fisherman used fishing lines of several different gauges and recorded the number of fish caught on each.

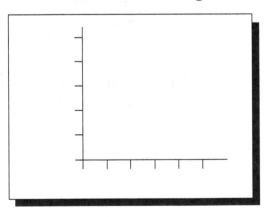

Gauge of Line (test pounds)	Average Number of Fish Caught
6	1
8	5
10	12
12	20
15	37
20	22

4. SELF CHECK

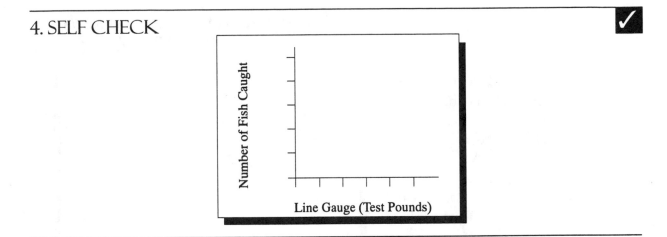

> 5. *INVESTIGATION:* A study was conducted to see if the number of surfers on the beach was affected by the average height of the waves.

Average Height of Waves (m)	Number of Surfers
1	13
2	23
3	56
4	31

How Is the Number of Surfers on the Beach Affected by the Height of the Waves?

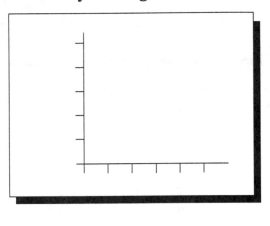

5. SELF-CHECK

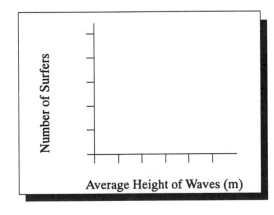

How Does Depth of Collection Affect the Density of Rocks?

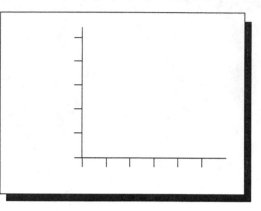

➤ 6. *INVESTIGATION:* Rocks from several different depths in a mine were collected. The density of each rock was recorded.

Depth of Collection (m)	Density of Rocks (g/cm³)
0	2.2
30	2.0
120	2.7
600	3.5
3000	4.0

6. SELF-CHECK ✓

Remember, by convention, the manipulated variable is written in the left column of a data table. Read the description of an investigation to decide which variable has been manipulated.

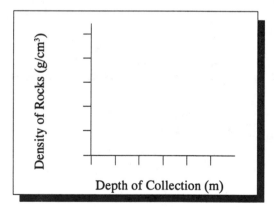

Skill 2 — Determining Interval Scales for Each Axis

Refer to this data table for the following example:

7	
12	
22	
37	
46	
55	

Finding the right scale for numbering the axes of a graph can be challenging. Trial and error is one approach that may help find the right scale for each axis. However, most students are not very good with this approach; they need a little structure to guide them. An easy way to find a good scale to fit the data consists of the following steps:

STEPS	EXAMPLE

STEPS

1. Find the range of the data to be graphed by subtracting the smallest value from the largest value.

2. Divide this difference by the number of intervals you want. If you want about 5 intervals, divide by 5. This usually results in a scale with 5 to 7 intervals. Too many intervals crowd a graph, while too few make it difficult to plot data points.

3. Using 9.6 to make intervals would be awkward, so to make the job easier round 9.6 to an easy counting number like 10. Good counting numbers are usually multiples of 5, like 10 or 20, or smaller numbers like 2 and 4.

4. Use this rounded number (10 in this example) to mark off the intervals along the axis. Begin with a multiple of 10 that is less than the smallest value to be plotted (7 in this example) and continue until you have reached or exceeded the largest value to be plotted. Numbering for both axes always begins at the origin of the axes. These steps result in a scale that uses the entire graph area.

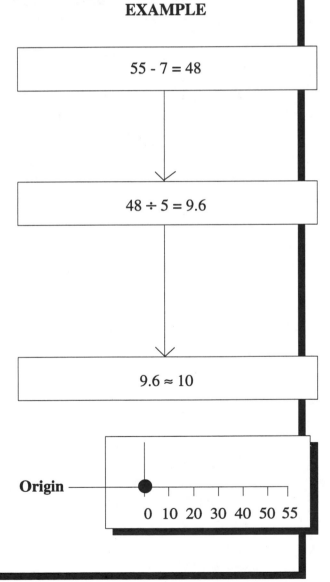

$$55 - 7 = 48$$

$$48 \div 5 = 9.6$$

$$9.6 \approx 10$$

Origin

0 10 20 30 40 50 55

© Rezba, Sprague, Fiel, Funk, Okey, Jaus, **Learning and Assessing Science Process Skills**. Kendall/Hunt 1995.

For questions 7-12, refer to the following table. For each question, determine if the interval scale is correct.

Average Cost of Watch ($)	Error per Month (minutes)
249	2
225	4
220	5
200	6
124	9
110	10

➤ 7.

A. Scale is OK ❑
B. Size of intervals not equal ❑
C. Too many intervals ❑
D. Too few intervals ❑
E. Starts with too small an interval ❑

0 50 100 150 200 250
Average Cost of Watch ($)

7. SELF-CHECK ✓

E, by starting with 0 only about half of the available space on the graph is used. The first number on the axis should be the smallest number to be graphed or some number only slightly smaller. It is not necessary to start with 0. For instance, in this problem the first number should have been 100 or some number smaller than 110.

➤ 8.

A. Scale is OK ❑
B. Size of intervals not equal ❑
C. Too many intervals ❑
D. Too few intervals ❑
E. Starts with too small an interval ❑

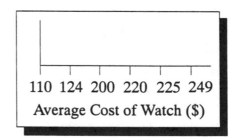

110 124 200 220 225 249
Average Cost of Watch ($)

8. SELF-CHECK ✓

B, the difference between 110 and 124 is not the same as the difference between 124 and 200. The size of the intervals must be equal.

➤ 9.
    A. Scale is OK ❑
    B. Size of intervals not equal ❑
    C. Too many intervals ❑
    D. Too few intervals ❑
    E. Starts with too small an interval ❑

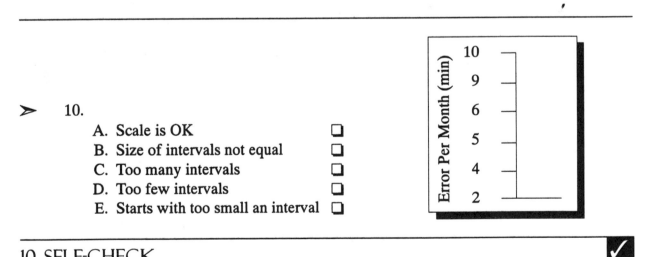

9. SELF-CHECK ✓

A, scale is correct. It begins with an interval just below the smallest value and continues with equal intervals of 30 to accommodate all the data.

➤ 10.
    A. Scale is OK ❑
    B. Size of intervals not equal ❑
    C. Too many intervals ❑
    D. Too few intervals ❑
    E. Starts with too small an interval ❑

10. SELF-CHECK ✓

B, intervals are not equal.

➤   11.

A. Scale is OK ☐
B. Size of intervals not equal ☐
C. Too many intervals ☐
D. Too few intervals ☐
E. Starts with too small an interval ☐

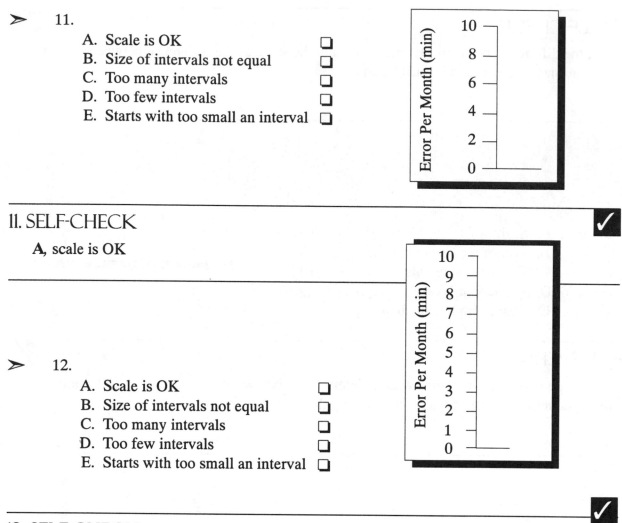

11. SELF-CHECK ✓

A, scale is OK

➤   12.

A. Scale is OK ☐
B. Size of intervals not equal ☐
C. Too many intervals ☐
D. Too few intervals ☐
E. Starts with too small an interval ☐

✓

12. SELF-CHECK

C, there are too many intervals cluttering the scale; 5 or 6 intervals would be better.

You have just had some practice in distinguishing between proper and improper numerical scales. In the following exercise, you will use your skills to determine scales of your own.

➤ 13. Mark off the horizontal axis of the following graph outline into five or so equal line segments. Use a series of small marks along the axis. Mark the vertical axis into about five equal line segments also.

How Does the Height of a Plant Affect the Number of Leaves?

Height of Plant (cm)	Number of Leaves
36	68
42	73
47	90
50	180
53	116
57	216

Examine the table of data above. Determine the range of the data for *Height of Plant (cm)* by subtracting the smallest value from the largest value. The difference between 36 and 57 is 21. Divide by the number of desired intervals (we chose 5) to determine the size of the intervals. When 21 is divided by 5, the result is 4.2. Round 4.2 to a convenient counting number such as 5. The smallest number you must graph is 36. Begin labeling the axis with a multiple of 5 that is less than 36 (35).

➤ 14. Label each mark on the horizontal axis of #13 in intervals of 5. Begin the interval scale with 35.

---

14. SELF-CHECK ✓

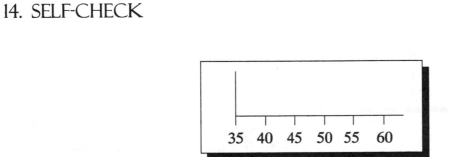

Refer to the data table in #13 again. What is the difference between the largest and smallest value for the responding variable *Number of Leaves*? 216 - 68 = 148. Divide this range by 5 to determine the size of the interval, 148 ÷ 5 = 29.6. Counting by 30's is easier than by 29's, so round the number to 30. Begin the scale with a multiple of 30 (60) that is less than the smallest value to be graphed (68).

➤   15. Starting with 60, label the vertical axis of #13 with intervals of 30.

15. SELF-CHECK

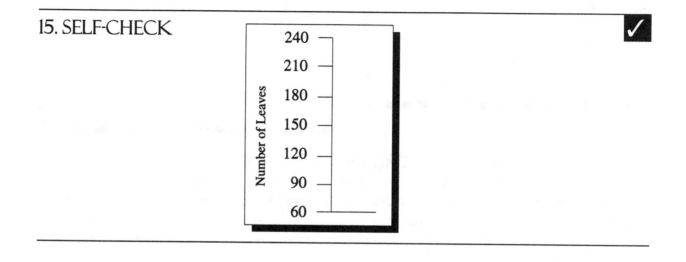

➤   16. Here is a second problem. Label the axes with proper numerical scales for graphing the data in the table below. Remember to mark off each axis with about 5 marks and then devise an appropriate numerical scale for each.

The Effect of Hatching Time on Number of Flies Hatched

Time from Start of Hatching (hours)	Number of Flies Hatched
9	25
12	153
14	269
15	617
18	1245

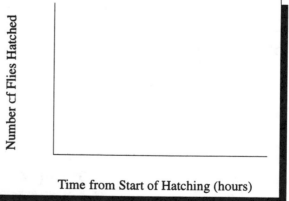

16. SELF-CHECK ✓

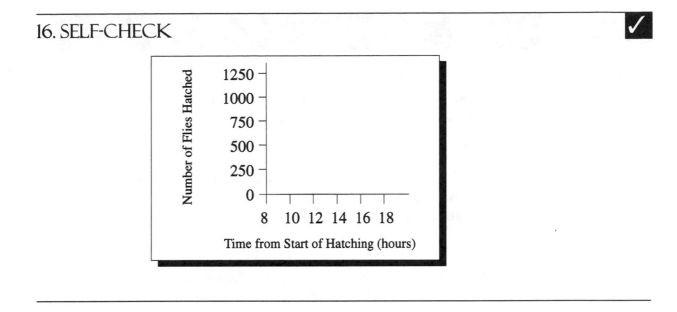

➤ 17. Try one more problem on determining an interval scale. Label the axes with appropriate scales for graphing these data.

**The Effect of Mower Width
on Mowing Time**

Width of Mower (cm)	Time to Mow Field (min)
37	240
43	210
110	175
125	160
180	143

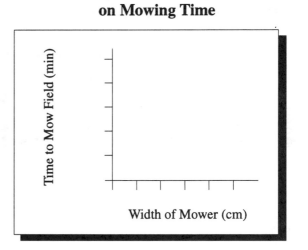

17. SELF-CHECK ✓

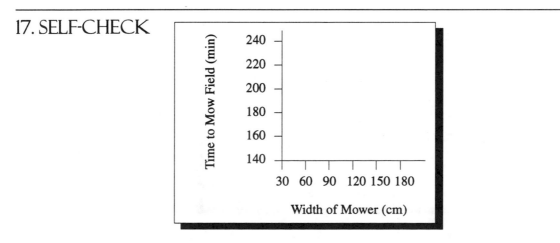

Skill 3 Plotting Data Pairs as Points on a Graph

You should now be ready for the third skill you need to construct a graph. After you have labeled the variables along the X and Y axes and have determined appropriate scales for each axis, you are ready to plot points for data pairs in a table of data.

➤ 18. Refer to the following table of data. The first data pair is (8,6). Locate 8 on the horizontal axis and 6 on the vertical axis. Imagine a vertical line drawn straight up from the 8 and a horizontal line drawn straight across from the 6. Where these two imaginary lines intersect is a point representing that data pair.

Look at the second pair of data in the table (10, 15). Sight imaginary lines straight up from 10 and straight across from 15. The point at which these imaginary lines intersect is a point representing the data pair, (10, 15).

Time of Day (Clock Time)	Number of Oxygen Bubbles Per Minute
8	6
10	15
12	27
2	19
4	5

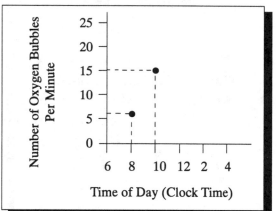

The Effect of Time of Day on Number of Oxygen Bubbles

18. SELF-CHECK ✓

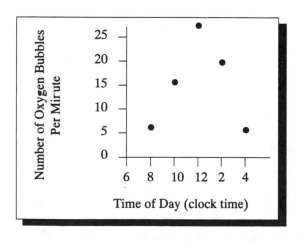

➤ 19. Here is another table of data. The positions of the first two data pairs from this table are already plotted on the graph. You plot the other four data pairs.

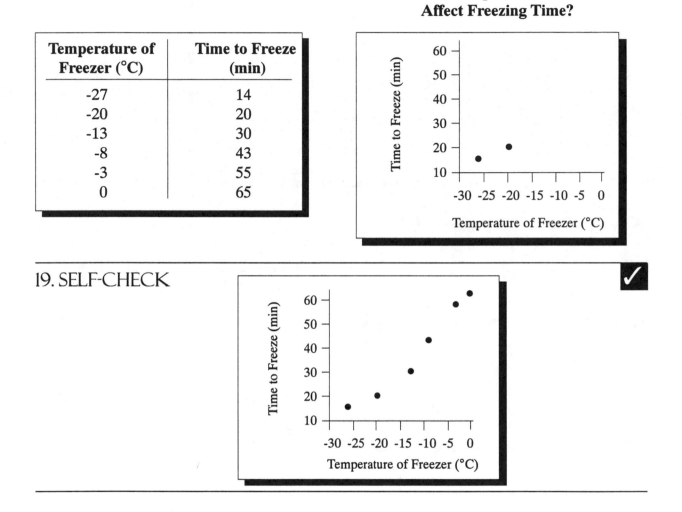

How Does the Temperature of the Freezer Affect Freezing Time?

Temperature of Freezer (°C)	Time to Freeze (min)
-27	14
-20	20
-13	30
-8	43
-3	55
0	65

19. SELF-CHECK ✓

➤ 20. Plot the data pairs on the graph using this table of data.

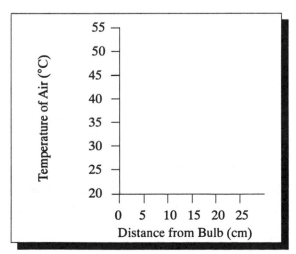

How Does Distance from a Light Bulb Affect Air Temperature?

Distance from Bulb (cm)	Temperature of Air (°C)
5	55
10	40
15	31
20	28
25	25

20. SELF-CHECK ✓

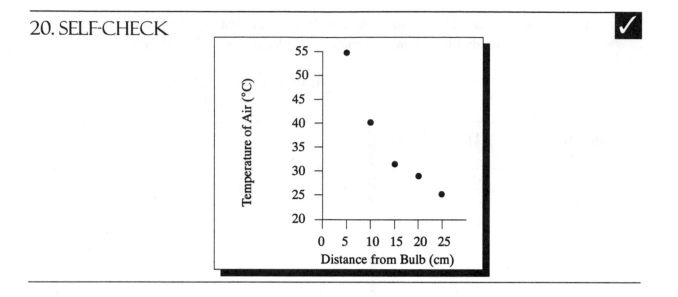

➤ 21. Plot the data points for this table on the graph.

How Does the Date Affect the Number of Library Books Checked Out?

Date in November	Number of Books Checked Out
8	675
13	353
15	430
20	270

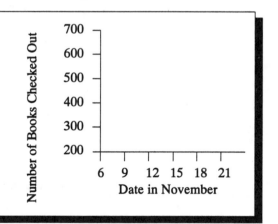

21. SELF-CHECK ✓

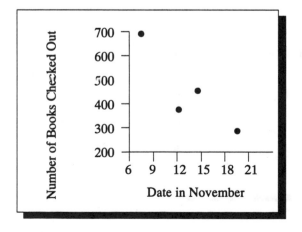

You have just practiced the third skill you will need in order to construct a graph. All you have to do now is put together the skills you have learned and you will be able to construct a graph.

Remember, for a graph to be properly constructed you must be able to do three things

1. Write labels for the variables along the correct axis.
2. Determine an appropriate interval scale for each axis.
3. Plot each data pair as a data point on the graph.

In the last part of this chapter you will examine several graphs to see if they have been constructed correctly. Then when given a description of an investigation and a table of data, you will be expected to construct your own graphs.

An investigation and a table of data follow. Two graphs of the data are shown. For each graph determine if it is constructed correctly. If you think the graph is not correct, check the reason it is incorrect.

INVESTIGATION: The temperature of the air was measured at several times during the day.

Time of Day (clock time)	Temperature of Air (°C)
8	7
10	10
12	17
2	14
4	9

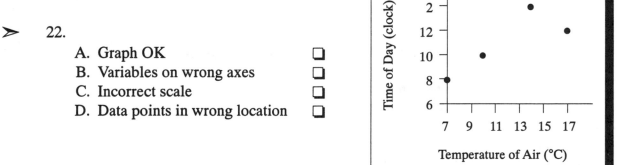

The Effect of Time of Day on Temperature of Air

➤ 22.
   A. Graph OK ❏
   B. Variables on wrong axes ❏
   C. Incorrect scale ❏
   D. Data points in wrong location ❏

---

22. SELF-CHECK ✓

B, the axes are labeled with the variables reversed. The *time of day* is the manipulated variable so it should be on the horizontal axes.

➤ 23.

A. Graph OK ☐
B. Variables on wrong axes ☐
C. Incorrect scale ☐
D. Data points in wrong location ☐

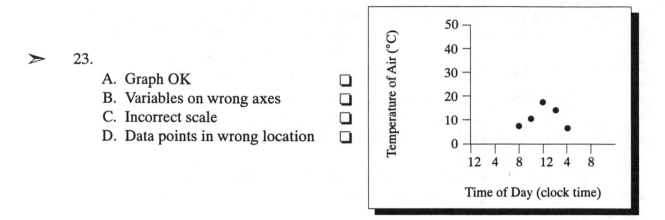

23. SELF-CHECK ✓

C, the numerical scales on both axes are too large. Notice that the data points are all in a small lump. With appropriate scales the points should be distributed throughout the graph. Now try putting all three skills together.

➤ 24. Construct a graph of data in this table.

How Does Car Speed Affect Gasoline Consumption?

INVESTIGATION: The number of kilometers per liter of gasoline was measured for cars traveling at different speeds.

Speed of Car (km/h)	Kilometers per Liter
20	6.0
25	5.5
30	4.0
40	3.5

24. SELF-CHECK ✓

How Does Car Speed Affect Gasoline Consumption?

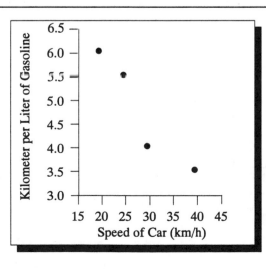

➤ 25. Try another problem. Construct a graph for the data from this investigation.

INVESTIGATION: The average weight of ten pumpkins growing in a patch was determined at different times after planting.

Time after Planting (weeks)	Average Weight of Pumpkins (kg)
2	0
7	0
9	1
12	9
16	15
18	22

The Effect of Planting Time on Pumpkin Weight

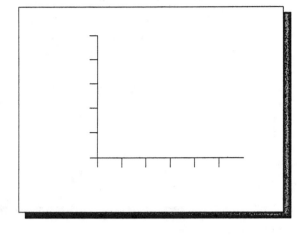

25. SELF-CHECK ✓

The Effect of Planting Time on Pumpkin Weight

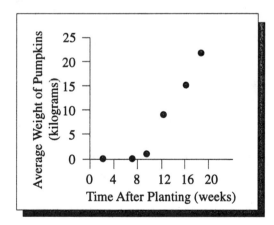

➤ 26.

INVESTIGATION: A box was dropped from an airplane and the distance it had fallen was measured after various lengths of time.

The Effect of Time on Distance Fallen

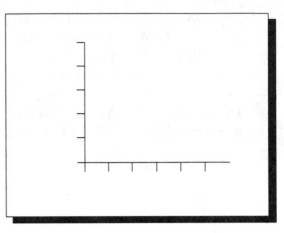

Time (sec)	Distance Fallen (m)
1	5
2	20
3	45
4	80
5	125

26. SELF-CHECK ✓

The Effect of Time on Distance Fallen

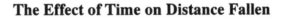
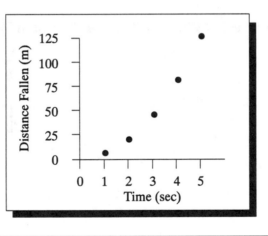

You have just learned to construct a graph of the data in a table. This is a valuable skill you will use later in this program and, hopefully, with your students. In Chapter 10 you will learn to write a summary description of a graph.

Now take the Self-assessment for Chapter 9.

Self-Assessment Constructing a Graph

> 1. a. Write labels for the variables from this investigation along the appropriate axes of the graph.

   INVESTIGATION: The temperature of the water was varied in several containers to see if the time to evaporate the water was affected.

   b. What is the rule used to label the axes of a graph?

> 2. Label the axes with appropriate interval scales for graphing these data. DO NOT locate the points.

   INVESTIGATION: The temperature of the air was measured on different days to see if it affected the number of swimmers on the beach.

Temperature of Air (°C)	Number of Swimmers
12	30
19	80
20	225
26	450
31	475

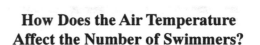

How Does the Air Temperature Affect the Number of Swimmers?

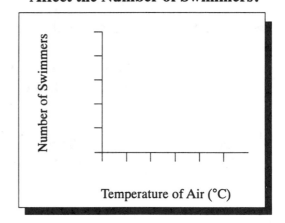

➤ 3. Plot the data pairs from this investigation on the graph.

INVESTIGATION: The number of letters that could be correctly identified on an eye chart at different distances was investigated.

Distance of Eye from Chart (m)	Number of Letters Identified
1	18
2	22
3	34
4	30
5	26

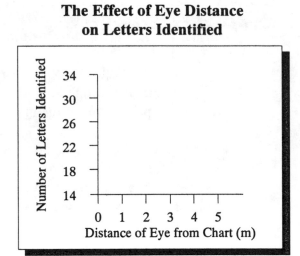

The Effect of Eye Distance on Letters Identified

➤ 4. Construct a graph of the data in this table.

INVESTIGATION: Ice cubes of different sizes are melted in a pan of water.

Mass of ice Cubes (g)	Time to Melt (min)
35	2
45	3
52	5
61	9
70	11

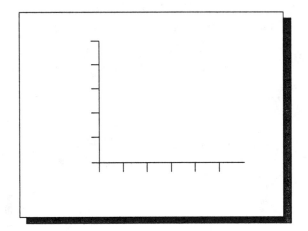

SELF-ASSESSMENT ANSWERS

➤ 1. a.

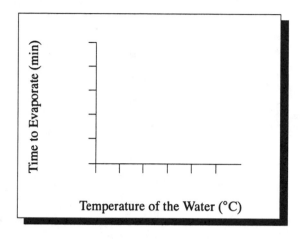

b. Rule: The manipulated variable is written along the horizontal axis.

➤ 2.

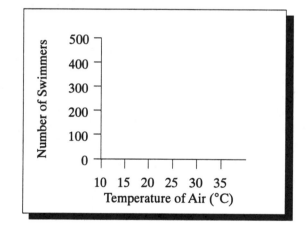

➤   3.

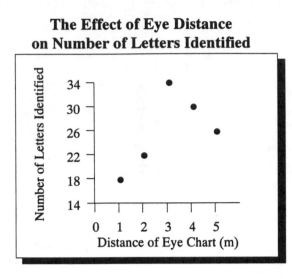

The Effect of Eye Distance
on Number of Letters Identified

➤   4.

How Does the Mass of Ice Cubes Affect Melting Time?

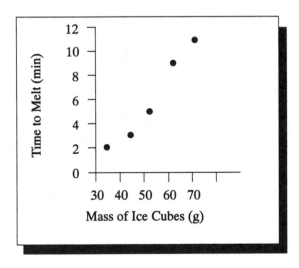

To be correct:
a. the variables must be written along the axes as shown
b. the interval scales should be about the same as these - if you numbered from 0 to 70 by 10's on the horizontal scale, it is wrong.

ASSESSING FOR SUCCESS: TEACHER RATING SHEET

Name: _____ Date: _____

Graph Title: _____

CRITERIA	TEACHER RATING	
GRAPHING SKILLS	**Possible Points**	**Earned**
Does the title communicate the MV and RV?	5	
Is the MV on the X axis?	10	
Are the label/units given for the MV?	10	
Is the scale on the X axis appropriate to represent the values of the MV?	10	
Is the RV on the Y axis?	10	
Are the label/units given for the RV?	10	
Is the scale on the Y axis appropriate to represent the values of the RV?	10	
Are the data correctly plotted ?	15	
Is the line-of-best fit appropriate?	10	
Is the graph done neatly?	10	

Describing Relationships Between Variables

Sight

Smell

Sound

Taste

Touch

PURPOSE

In the last two chapters you learned to organize data in a table and to construct a graph. One other skill associated with graphing needs to be learned - the skill of interpreting a graph.

You might think of a graph as a coded message; it means a great deal to the person who understands the code but not much to anyone else.

OBJECTIVES

After studying the information in this chapter you should be able to:

1. Draw a best-fit line when given a graph.
2. Describe in writing the relationship between variables on a graph.

Approximate time for completion: 45 minutes

First you will learn to draw a line of best-fit. The rules for constructing a best-fit line for a set of data points on a graph are:

1. The line should be a straight line or a smooth curve.
2. All points should lie either on the line or very near to the line.
3. There should be approximately equal number of data points on either side of the line.

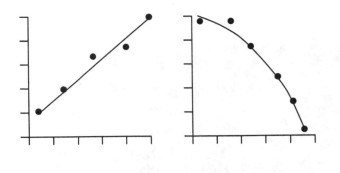

For example, examine the lines drawn on the graphs shown here. They are examples of best-fit lines.

Notice that in both cases the best-fit line either passes directly through or very near all the data points. If the points do not lie directly on the line, there are approximately equal numbers of points on either side of the best-fit line.

Shown below are several graphs with lines drawn through the data points. You are to decide whether it is a best-fit line. If you decide the line is not the "best-fit" line, check the reason.

➤ 1.

A. Line is best-fit   ❏
B. Should be curved   ❏
C. Should be straight   ❏
D. Too many data points on one side   ❏
E. Curve not smooth   ❏

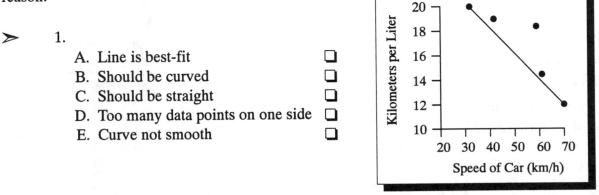

1. SELF-CHECK

D, the line should be moved a little toward the upper right. You may have checked **B** also. A curved line could be used here.

➤ 2.

A. Line is best-fit   ❏
B. Should be curved   ❏
C. Should be straight   ❏
D. Too many data points on one side   ❏
E. Curve not smooth   ❏

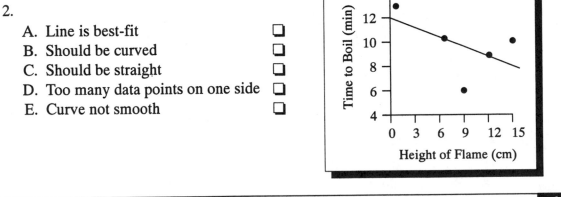

2. SELF-CHECK

B, some of the points are a long way from the straight line. A curved line drawn in the shape of a "U" would fit better.

➤ 3.
    A. Line is best-fit ❑
    B. Should be curved ❑
    C. Should be straight ❑
    D. Too many data points on one side ❑
    E. Curve not smooth ❑

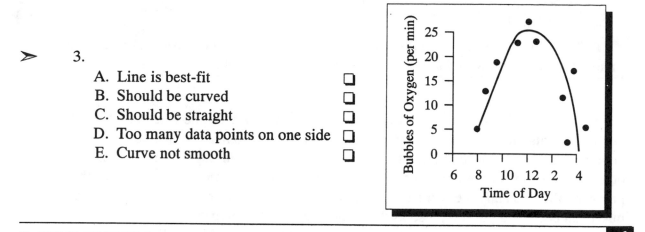

3. SELF-CHECK

A, notice that the line seems to *average* the points. Some are above the line and some are below.

➤ 4.
    A. Line is best-fit ❑
    B. Should be curved ❑
    C. Should be straight ❑
    D. Too many data points on one side ❑
    E. Curve not smooth ❑

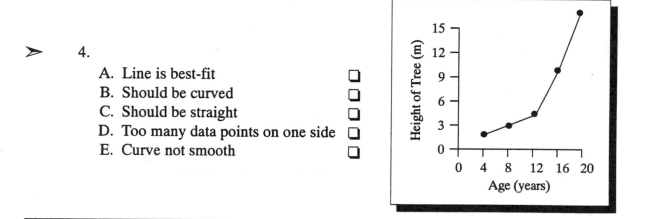

4. SELF-CHECK

E, a best-fit line never connects point to point in straight line segments. If it curves, it should be a smooth curve. For the graph above, a smooth curve shaped like a "J" would probably be the line of best-fit.

➤ 5.
    A. Line is best-fit ❑
    B. Should be curved ❑
    C. Should be straight ❑
    D. Too many data points on one side ❑
    E. Curve not smooth ❑

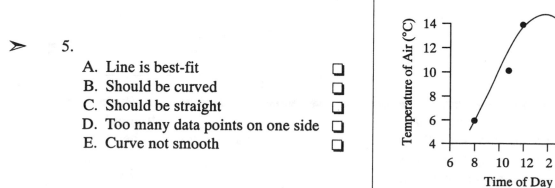

5. SELF-CHECK

D. There are more points on the inside of the curve than on the outside. The curved line should probably be lowered a little in order to average the points on the graph.

Now try drawing best-fit lines for the graphs below. First, decide whether a straight or curved line fits the points best. Then draw in the line. Try to make your lines *average out* the points on the graph. A good line of best-fit will usually pass through a few points, be above others, and below still others. Such a line then represents the numerical average of the data points on the line and on either side of the line. As you practice drawing best-fit lines, keep in mind that each data point represents the numerical relationship between the variables being investigated. The best-fit line provides a *picture* of the relationship of all the data points to each other.

➤ 6. Draw best-fit lines for these points.

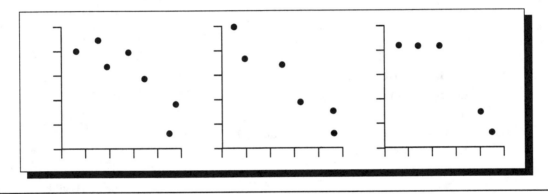

6. SELF-CHECK

Here are the lines we drew. These is some room for differences but your lines should be similar to these.

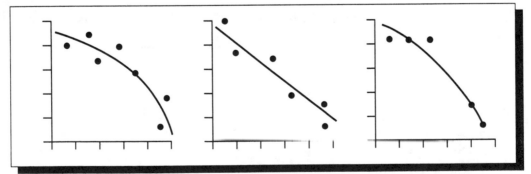

| Some of the points are quite far from the line, but equal numbers are found on either side of the line. | All the points do not have to be equal distances from the line. | A straight line would have resulted in some points being a great distance from the line. |

7. Now try again. Draw best-fit lines for the points on these graphs.

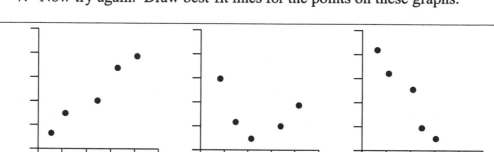

Hint: Holding a graph at arm's length for a moment may help you visualize the general shape of an appropriate best-fit line.

7. SELF CHECK ✓

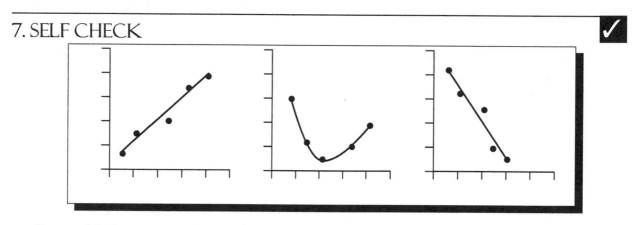

Be careful about *averaging out* the points. The number of points not on the line and their distance from the line should be approximately equal along both sides of the line.

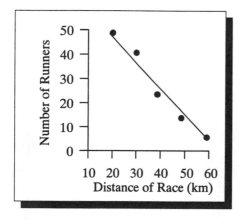

You have just learned how to draw a line of best-fit. Now all you have to learn is to write a statement that describes the relationship between the variables on a graph.

For example, if you were given the graph shown here, you should be able to write a statement that summarizes the relationship between the manipulated and responding variables: *The number of runners decreases steadily as the distance of the race increases.*

A procedure for describing the relationship between variables on a graph is as follows:

Tell what happens to the responding variable as the manipulated variable changes. A statement of relationship might read like this:

The temperature of water increases as the length of time it is heated increases.

➤ 8. Examine the following graph. Follow the line on the graph as it moves from left to right. Does the value of the responding variable increase or decrease?

INVESTIGATION: Ropes of different diameter are tested to see how much they will hold before breaking.

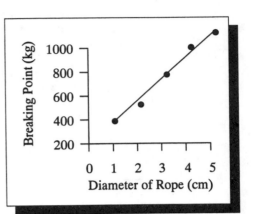

8. SELF CHECK

Increase

➤ 9. Using the procedure given earlier, write the name of the responding variable and how it changes.

9. SELF-CHECK

The breaking point increases.

➤ 10. Now write *as* followed by the manipulated variable and how it changes.

10. SELF-CHECK

as the diameter of the rope increases.

You have now described the relationship between the two variables represented on the graph: ***The breaking point increases as the diameter of the rope increases.***

Try another one.

INVESTIGATION: The number of letters recognized on an eye chart was measured to see if it was affected by distance.

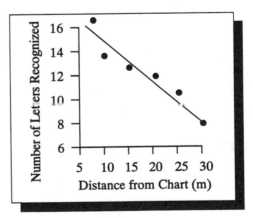

➤ 11. What happens to the responding variable as the graph moves toward the right?

11. SELF-CHECK ✓

It decreases

➤ 12. Write the name of the responding variable and how it changes.

12. SELF-CHECK ✓

The number of letters recognized decreases

➤ 13. Follow this with "as" and the manipulated variable and how it changes.

13. SELF-CHECK ✓

as the distance from the chart increases.

The complete statement of relationship is: **The number of letters recognized decreases as the distance from the chart increases.**

Two graphs are given here. Write a statement of the relationship between the variables for each graph. Remember to use the procedure on page 177.

➤ 14. _____
_____
_____

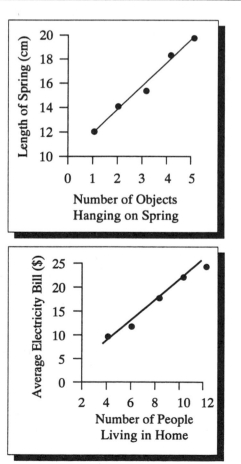

Number of Objects Hanging on Spring

14. SELF-CHECK ✓

The length of a spring increases as the number of objects hanging from it increases.

➤ 15. _____
_____
_____

Number of People Living in Home

15. SELF-CHECK ✓

The average electricity bill increased as the number of people living in the home increased.

The steps for describing the relationship between the variables on a curved line graph are as follows:

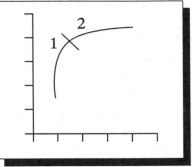

1. Describe the relationship in two sentences.
2. First describe the relationship until the curve changes direction.
3. Then tell what the relationship is for the rest of the graph.

Examine the curve at the right. The first sentence in a description should describe the section marked "1". The second sentence should describe the section marked "2."

A change in the direction of a line indicates a change in the relationships between the variables.

Examine the graph given below.

INVESTIGATION: An ice cube is placed in a glass of water and the temperature of the water is measured every few minutes.

➤ 16. Place a mark across the best-fit line about where it first bends.

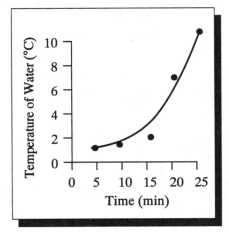

16. SELF-CHECK ✓

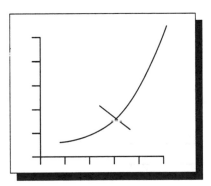

➤ 17. In one sentence describe what happens on the graph up until the mark you drew.

17. SELF-CHECK

The temperature of the water increased slowly as time passed.
Your statement does not have to be exactly the same as this but it should be similar.

➤ 18. Now describe what happens on the graph above the mark you drew.

18. SELF-CHECK

After 15 minutes the temperature of the water increased rapidly as time passed.

Together these two sentences should describe the entire graph.
Now try another.

> *INVESTIGATION:* Tomato plants were grown at several different temperatures. The average number of tomatoes produced by each plant was measured.

➤ 19. Mark where the best-fit line changes direction.

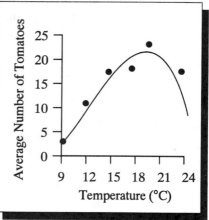

19. SELF-CHECK

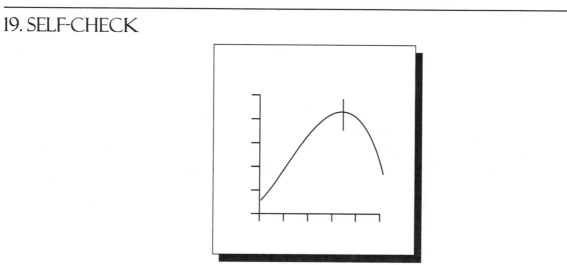

➤   20. Describe what happens on the graph up until the mark you drew.

20. SELF-CHECK

The average number of tomatoes produced increased rapidly until a temperature of 19 °C was reached.

➤   21. Now describe what happens on the graph after the mark you drew.

21. SELF-CHECK

Above 19 °C the average number of tomatoes produced declined rapidly.

INVESTIGATION: A pan of water is heated over a burner and the temperature is recorded every two minutes.

➤   22. Write a statement of the relationship between the variables shown on the graph above.

_____
_____
_____
_____

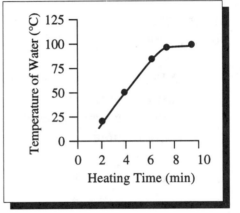

22. SELF-CHECK

The temperature of water steadily increases when heated for 7 minutes. After that the temperature stays about the same even though heating continues.

In the materials just completed you have learned to draw a best-fit line and write a statement of the relationship between the variables on a graph. In Chapter 9 you learned to construct a graph. Now you are ready to try practice problems in which you put all three of these skills together. However, before you do it on your own you will critique two problems in which someone else has made a graph and described it.

Two descriptions of investigations and the data collected from each are given below. Also given are a graph of the data, a best-fit line, and a statement of the relationship between the variables. You are to describe whether each has been properly prepared. If a section has not been correctly presented, check what part is wrong.

INVESTIGATION: The sea otters in a sheltered lagoon were counted over a number of years. These are the recorded data.

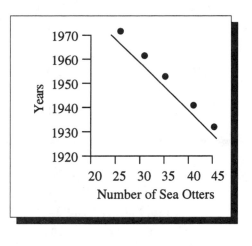

Year	Number of Sea Otters
1932	46
1940	42
1952	35
1962	30
1972	26

The number of sea otters in this location has been steadily decreasing since 1932.

➤   23.

Variables on wrong axis	❑
Graph OK	❑
Numerical scale wrong	❑
Number pairs in wrong position	❑
Statement OK	❑
Should be two sentences	❑
Does not include both variables	❑
Best-fit line OK	❑
Wrong shape line	❑
Line does not average points	❑

23. SELF CHECK   ✓

Variables on wrong axis.
Statement is OK.
Line does not average points (all the points are above the line).

INVESTIGATION: The average number of hits on a target in an archery contest was measured at different distances from the target.

Distance from Target (m)	Average Number of Hits
15	23
35	22
50	20
75	15
90	4

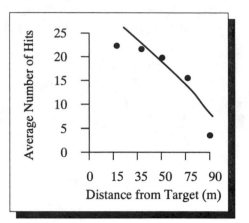

➤ 24.

Graph OK ❑
Variables on wrong axis ❑
Numerical scale wrong ❑
Number pairs in wrong position ❑

Statement OK ❑
Should be two sentences ❑
Does not include both variables ❑

Best-fit line OK ❑
Wrong shape line ❑
Line does not average points ❑

The average number of hits decreased steadily as the distance from the target increased

24. SELF CHECK ✓

Numerical scale wrong (intervals on horizontal axis are not of equal value).
Should be two sentences in the description.
Wrong shape best-fit line (should be a curved line).

Now that you have critiqued some problems, you should be ready to put all the graphing skills you have learned to use.

Two descriptions of investigations and a table of data for each are presented. For each, do the following:

a. Construct a graph of the data.
b. Draw a best-fit line.
c. Write a description of the relationship between the variables.

➤ 25. *INVESTIGATION:* A potato is cut in two and allowed to dry in the sun. The weight of the potato is measured as the days pass.

Elapsed Time (days)	Weight of Potato (grams)
1	330
7	300
12	180
14	150
21	160
26	120

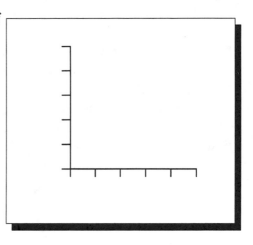

25. SELF-CHECK ✓

As the days pass, the weight of the potato drops rapidly until the 15th day. After that the weight loss is very slow and seems to be stopping.

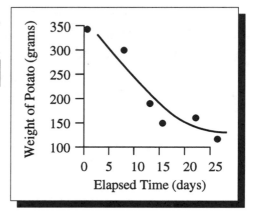

➤ 26. *INVESTIGATION:* A fire chief is doing an analysis of his men at work. He measures the average time it takes a fireman to climb ladders of different lengths.

Length of Ladder (m)	Time to Climb (sec)
1	2
2	5
3	8
8	18
12	22
15	53

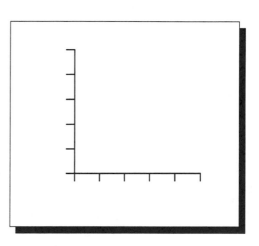

26. SELF-CHECK ✓

Note: Either a curved or straight best-fit line could be used here. As the length of the ladder increases, the time to climb it steadily increases. If you drew a curved line of best-fit, you may want to add a second sentence like this one. Above lengths of 12 meters, the time to climb increases more rapidly.

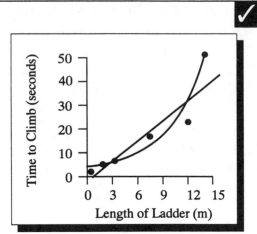

If you were successful on the last two problems, you have learned some important skills that are helpful in solving science problems. In the part of the program you will study next, you will learn to acquire and process your own data.

Now take the Self-assessment for Chapter 10.

Self-Assessment · Describing Relationships between Variables

DESCRIBING RELATIONSHIPS BETWEEN VARIABLES

➤ 1. A description of an investigation and a table of data are given here.

a. Construct a graph.
b. Draw a best-fit line.
c. Write a statement of the relationship between the variables.

INVESTIGATION: An investigation was carried out to determine the relationship between the size of an automobile motor and the gasoline mileage.

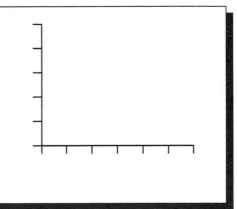

Size of Motor (horsepower)	Average Kilometers per Liter of Gasoline
47	7.0
100	5.0
140	4.0
193	3.5
227	3.0

➤ 2. Draw a best-fit line for the points on the graph.

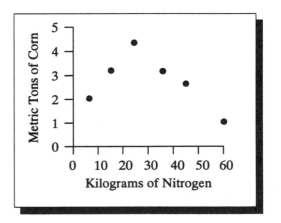

➤   3. Write a statement of the relationship between the variables shown on this graph.

   INVESTIGATION: A weather station kept a record for a ten year period of the average amount of rainfall during several months.

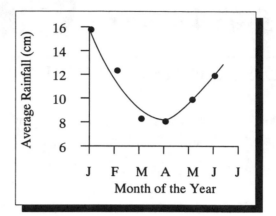

_____

_____

_____

➤   4. Write a statement of the relationship between the variables on this graph.

   INVESTIGATION: Some soldiers were tested to see if the number of kilometers they could hike in an hour was affected by the temperature.

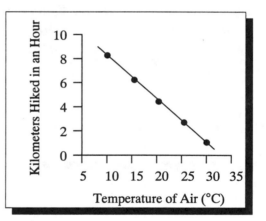

_____

_____

_____

SELF-ASSESSMENT ANSWERS

➤ 1.

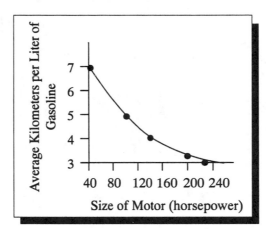

The number of kilometers per liter of gasoline decreases as the size of the motor increases. However, the decrease is slower for motors above 120 horsepower.

➤ 2.

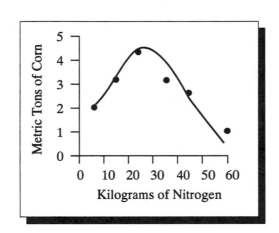

➤ 3. The average rainfall per month steadily decreases between January and April. The average rainfall per month steadily increases from April to June.

➤ 4. The number of kilometers marched in an hour by soldiers steadily decreases as the temperature of the air increases.

ASSESSING FOR SUCCESS: PERFORMANCE TASK

Preparation

Pre-assemble a small plastic bag of materials for each student or group of students doing the task at the same time. Use large washers, small fishing sinkers, or similar weights as the pendulum bob. Make 6 strings, each with a loop tied at one end and a paper clip at the other end. Open the paper clips slightly to serve as a hook for the pendulum bob. The string lengths from the top of the loop to end of the pendulum bob should be 10, 20, 30, 40, 50, and 60 centimeters. Have several rolls of masking tape available or distribute strips of tape to participating students. Groups of students may complete Part I of this task together but Part II should be completed individually.

Directions to the Student

Part I

1. Check your materials.

   ✓ pencil
   ✓ 6 strings with a paper clip tied to one end
   ✓ 1 large washer (or other weight)
   ✓ roll or strip of tape
   ✓ clock or watch

2. Tape your pencil to your desk so it hangs over the edge.
3. Hang the shortest string (10 cm) from the pencil and hook your washer on the paper clip to make a pendulum.
4. Hold the pendulum even with the top of your desk and release it.
5. Count the number of times the pendulum swings back to the side where you released it in 15 seconds. Record your data in the table.
6. Repeat steps 3 - 5 using each of the 5 remaining strings (20, 30, 40, 50, and 60 cm)

Part II

1. Construct a graph of the results of this activity.
2. Draw a best-fit line.
3. Write a statement of the relationship between the variables.

Length of Pendulum (cm)	Number of Swings Back
10	
20	
30	
40	
50	
50	
60	

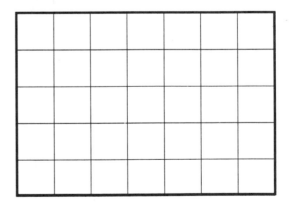

_____

_____

_____

Acquiring & Processing Your Own Data

Sight

Smell

Sound

Taste

Touch

PURPOSE

In the past three chapters you have been working with many data tables. However, the number pairs in these data tables were produced by someone else. In this chapter you will carry out several investigations and produce your own tables of data.

OBJECTIVES

After studying this chapter you should be able to:

1. Conduct an investigation and obtain a table of data.
2. Construct a graph of the data and a statement of the relationship between the variables. (This task was introduced to you in Chapters 9 and 10.)

Approximate time for completion: 60 minutes

MEASUREMENT SKILLS

In this chapter it is assumed that you know how to make measurements of mass, length, time temperature, force and volume. If you are not sure that you know how to make these measurements, you may wish to review Chapter 4.

GRAPH TITLES

You may have noticed that the graphs in Chapter 9 have titles. (The titles were deliberately omitted from the graphs in Chapter 10 so that you could concentrate solely on drawing best-fit lines.)

A graph title is very important because it communicates the purpose of the graph.

Learning to write a title for a graph is quite simple. A graph title may take this form:

The Effect of the _____MV_____ on the _____RV_____.

Or it may take the form of a question:

How Does the _____MV_____ affect the _____RV_____?

For example:

The Effect of a Basketball Player's Weight on How High He Can Jump
or
How Does the Weight of a Basketball Player Affect the Height He Can Jump?

From now on, when you are asked to construct a graph, give it a title by using either of these title formats. Each time you use a *self-check* also check the correctness of your graph title.

CONDUCTING AN INVESTIGATION

An experiment generally begins with a problem. Someone observes something occurring and wonders *Why?* For example, everyone knows that the old cliche, *a watched pot never boils*, is not true; however, it does raise an interesting problem. What determines the time it takes water to heat? Examine this problem a little closer. What are some of the variables that could affect the heating time of water?

_____   _____

_____   _____

_____   _____

SELF-CHECK ✔

Amount of water	Shape of container
Amount of dissolved material	Type of heat source
Height above sea level	

You may have come up with some that were different from these.

If the variable, *amount of dissolved material*, was selected for investigation, one could make this prediction: The time required to cause a change in temperature increases as the amount of material dissolved in water increases. The first exercise of data gathering will be to conduct the experiment proposed to test this hypothesis. The equipment you need and the directions you are to follow are given below.

✓ 4 pyrex beakers (100 mL)
✓ 1 graduated cylinder
✓ 1 spoon
✓ 1 hot plate
✓ 1 thermometer
✓ 1 timer
✓ sugar

Activity 1

Label the beakers 0, 1, 2, and 3. Measure 50 mL of water into each. Dissolve one spoon of sugar in beaker 1 and 2 spoons in beaker 2; and 3 in beaker 3. Place no sugar in beaker 0. Heat each beaker for three minutes. Record the change in temperature in the table below.

➤ 1. Table 11-1

Amount of Sugar (spoons)	Temperature Change (˚C)
0	
1	
2	
3	

SELF-CHECK ✓

Amount of Sugar (spoons)	Temperature Change (˚C)
0	38
1	35
2	32
3	31

These are data we collected.

2. Why did you use beaker 0?

_____

_____

To know that the sugar causes any change, it is necessary to know what happens when no amount of sugar is used. The beaker with no sugar serves as a standard of comparison to which all the other beakers with different amounts of sugar are compared. Most experiments include a standard of comparison, called a **control** or control group. In some experiments, such as this one, the control is called a **no treatment** control. In other experiments all groups receive a treatment (some amount 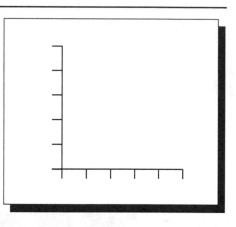 of the manipulated variable). The experimenter must then select one of the levels of the manipulated variable to serve as the control group. The level selected is usually the normal or typical case. For example, in an experiment on the effect of depth of seed on seed germination, the control might be the recommended or normal planting depth; other planting depths might be deeper or shallower than the control. This kind of control is called an **experimenter-selected** control.

3. Using the data obtained from your investigation, construct a graph and a *statement* about the relationship between the amount of dissolved sugar and the rate of temperature change. Be sure to give the graph a title.

_____

_____

SELF-CHECK ✓

Statement of Relationship

The temperature change decreases, as the amount of dissolved sugar increases. The addition of more than two spoons of sugar results in no increase in temperature change.

Your statement may be different from this one if your graph differs.

This is a graph of the data collected earlier. Yours may be different because the data you gathered were different from ours.

Does the Amount of Dissolved Sugar in Water Affect the Change in Temperature?

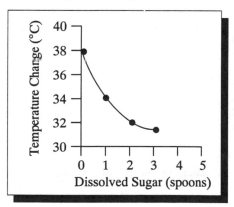

Here is another practice exercise:

Leticia and her cooperative learning team found an activity that might help them learn about their assigned project, space travel. The activity directions were:

1. Thread a plastic drinking straw onto a string that is long enough to reach from one side of the room to another.
2. Stretch the string tight between two sides of the room.
3. Tape one side of a plastic bag to the straw as illustrated below.
4. Blow up a balloon, hold it shut, and place the balloon in the bag.
5. Release the balloon and watch the balloon travel along the string—like a rocket.

Activity 2

⇨ *Obtain* the following materials and set-up the activity as Leticia and her team would.

You will need:

✓ 1 plastic drinking straw
✓ long piece of string
✓ 1 balloon
✓ masking tape
✓ 1 gallon size plastic bag

Follow the directions given for the activity. Conduct several trials to find out how the "rocket" works.

Leticia and her team wanted to turn this activity into an experiment by changing one variable and measuring the response. They chose the number of breaths blown into the balloon as their manipulated variable and the number of meters it traveled along the string as their responding variable. Their prediction was: As the number of breaths blown into the balloon are increased, the balloon will travel farther along the string.

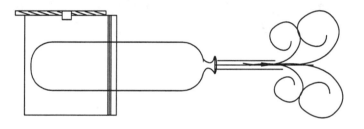

	Trials		
	1	2	3

Use your balloon rocket to test their prediction. Try different numbers of breaths (at least four) and conduct 3 repeated trials. Record your data in the following table and calculate an average distance for each breath.

The table of data constructed by Leticia's team follows. Your data will be different, but your table headings should be similar.

Table 11-2

How Does the Number of Breaths Affect the Distance Traveled by a Balloon Rocket?

Number of Breaths	Distance Traveled (m) Trials			Average Distance Traveled (m)
	1	**2**	**3**	
1	1.7	2.0	2.3	2.0
2	5.4	6.0	5.6	5.4
3	9.3	9.1	8.2	8.9
4	10.5	9.2	10.3	10.0

4. What other variables might affect how far the rocket travels?

_____

_____

SELF-CHECK ✓

Using different kinds (shape, diameter, length, composition, age) of balloons or even different balloons of the same kind may affect the distance traveled. If you used the same balloon, keep in mind that it has been stretched each time it was used for the previous launch. The amount of air in each breath should be the same and the way the rocket is launched should be the same. Were you able to think of some additional variables . . . such as the straw (its length, weight, diameter, material it is made of) . . . and the string (its angle, tautness, composition?)

5. Using the data obtained from your investigation, construct a graph and a statement about the relationship between the number of breaths blown into the balloon rocket and the average distance the rocket traveled. Be sure to give the graph a title.

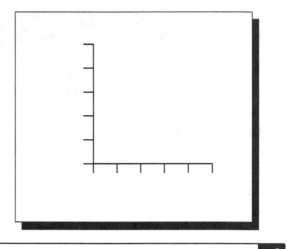

SELF-CHECK ✓

Leticia's team made the following graph. Your data will be different but your axis labels and graph title should be similar. Compare your statement of relationship with the one the team wrote.

Statement of Relationship
The distance the balloon traveled increased steadily as the number of breaths blown into the balloon increased.

Does the Number of Breaths Blown into a Balloon Rocket Affect the Distance the Rocket Travels?

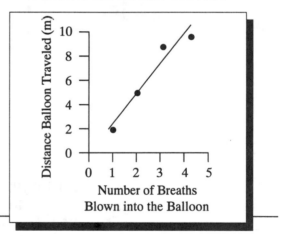

In your third practice problem you will use a siphon to empty a container. The data you gather are to be used to determine how the diameter of a siphon affects the time required to remove water from the container.

Assemble the equipment listed below and carry out the investigation as directed.

✓ 4 flexible hoses (each with a different inside diameter)
✓ 2 large containers
✓ 1 timer

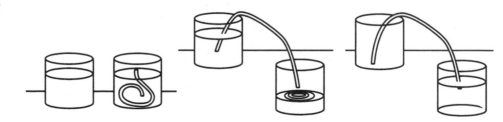

Activity 3

Fill one container to any level (as long as it is the same each time). Submerge the entire tube in the water and squeeze along the tube until all the air bubbles are out and the tube is full of water. Pinch one end closed and pull that end out and down the side until it is below the bottom of the beaker. Release the pinch and record the time required to empty the beaker. Follow the same procedure for each tube. If you have time, conduct repeated trials of each diameter tube. Record all your data in the data table below. Construct a graph and a statement of the relationship between the variables from the data obtained. (Don't forget the graph title.)

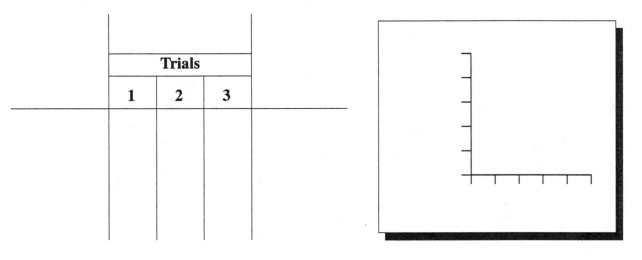

	Trials		
	1	2	3

SELF-CHECK

These are the data gathered during the investigation when we did it. You can compare your results with ours. Remember that the data you obtain and the graph you make may differ from ours and yet be correct. One difference might be in the amount of water - we used 300 mL.

Inside Diameter of Siphon (mm)	Time Required to Move 300 mL Water (sec)			Average Time Required to Move 300 mL Water (sec)
	Trials			
	1	2	3	
3	82	87	95	88
5	34	38	39	37
7	16	15	14	15
8	8	8	8	8

How Does the Diameter of a Tube Affect Siphoning Time?

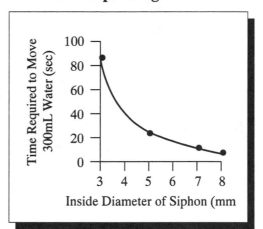

The time required to siphon water decreases rapidly as the inside diameter of the tube increased to about 7 mm. Above this, the time required to remove the water decreased very slowly.

You have now carried out a few investigations. In the following chapter you will examine parts of an investigation to help you design your own investigations.

Now take the Self-assessment.

Self-Assessment Acquiring and Processing Your Own Data

You are interested in determining the relationship between the amount of exercise an individual does and his pulse rate. To collect some data on this relationship carry out the following investigation.

Sit quietly for about 5 minutes and then count your number of heartbeats for 15 seconds. Now quickly step up onto a stool or stair step 5 times. Count your heartbeat for 15 seconds. Rest until your pulse rate returns to the resting rate. Quickly step up onto the stool 10 times. Count your heartbeat for 15 seconds. Repeat this procedure for 15, 20, and 25 step-ups.

Construct a data table, a graph, and a statement of the relationship between the variables for the data you gather.

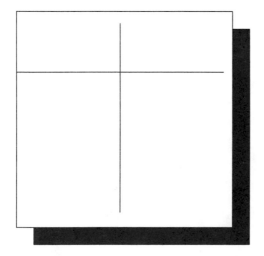

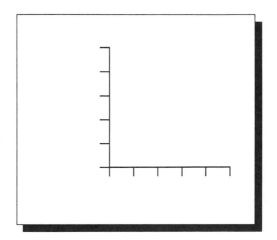

SELF ASSESSMENT ANSWERS

Here are the data gathered from an investigation we carried out. Your data will almost certainly be different.

Does the Number of Step-Ups Affect a Person's Heart Rate?

Numbers of Step-ups	Heartbeats (in 15 seconds)
0	20
5	26
10	28
15	32
20	36
25	39

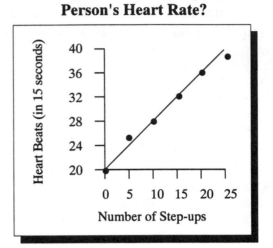

The pulse rate increases as the number of step-ups increase.

ASSESSING FOR SUCCESS: OPTICAL READER FOR TEACHER OBSERVATIONS

Learner Profile, an observational assessment tool, is an example of how technology is being applied to the classroom. By using a credit-card size optical reader and a list of bar-coded assessment items, teachers can instantly record observations of student behavior in any educational setting - the classroom, the science laboratory, school grounds, and field trips.

Barcodes are created for each student and for each item on desired lists of observations. Observations of a student's behavior are recorded by simply passing the optical reader over the student's barcode and then over the appropriate coded observations. Later, that information is loaded into a computer to analyze student progress, generate reports, and plan follow-up instruction.

The prototype system includes a small hand-held optical reader, an uploader to connect to a computer, and software to generate barcodes and to analyze and report data. Access to a Macintosh computer and a laser printer is required.

1. Learner Profile is produced by **Wings for Learning/Sunburst,** 1600 Green Hills Road, Scotts Valley, CA 95067-0002

STUDENTS

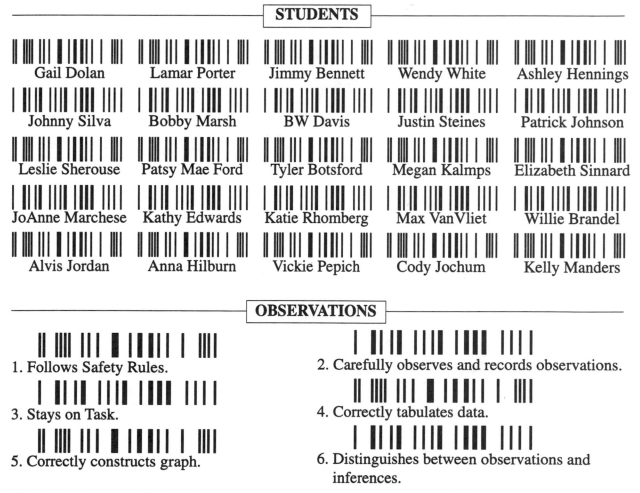

Gail Dolan	Lamar Porter	Jimmy Bennett	Wendy White	Ashley Hennings
Johnny Silva	Bobby Marsh	BW Davis	Justin Steines	Patrick Johnson
Leslie Sherouse	Patsy Mae Ford	Tyler Botsford	Megan Kalmps	Elizabeth Sinnard
JoAnne Marchese	Kathy Edwards	Katie Rhomberg	Max VanVliet	Willie Brandel
Alvis Jordan	Anna Hilburn	Vickie Pepich	Cody Jochum	Kelly Manders

OBSERVATIONS

1. Follows Safety Rules.

2. Carefully observes and records observations.

3. Stays on Task.

4. Correctly tabulates data.

5. Correctly constructs graph.

6. Distinguishes between observations and inferences.

©Rezba , Sprague, Fiel, Funk, Okey, & Jaus. **LEARNING AND ASSESSING SCIENCE PROCESS SKILLS,** Kendall/ Hunt 1995.

Analyzing Investigations

PURPOSE

Before you can design your own investigations, you need to learn to recognize the parts of a typical investigation. What are the variables under study? What hypothesis is being tested? These and other questions can be answered by analyzing an investigation.

OBJECTIVES

After studying this chapter you should be able to:

1. Identify the manipulated and responding variables and the constants in an experiment.
2. Identify the hypothesis being tested when supplied with a description of an investigation.

Approximate time for completion: 30 minutes

There are many factors which might affect the outcome of an experiment, factors that the experimenter may not be interested in at the moment. For example, suppose we wanted to test this hypothesis: ***The more salt dissolved in water, the higher the boiling temperature.***

To test this idea, we could heat a pot of water with a large quantity of salt dissolved, another pot of water with only a small amount of dissolved salt and another pot of water with no salt to serve as the control. However, any results obtained would be worthless if the amount of water used in each container was different or if the pots were made of different metals.

Perhaps now you are beginning to see one of the problems with investigations. We want to be able to say that the manipulated variable *and only the manipulated variable* affected the responding variable. We must make sure that any other factor that could affect the results is prevented from having an effect.

A factor that might affect an experiment but is kept from doing so is called a *constant*. An experiment with constants is said to be a controlled experiment. We could then define a constant as a factor that is prevented from affecting the outcome of the experiment.

What factors are kept the same in this experiment?

Place six cups of water in each of four identical coffee makers. Add one teaspoon of coffee to the first pot, two to the second, three to the third, and four to the last. Brew each batch of coffee for ten minutes.

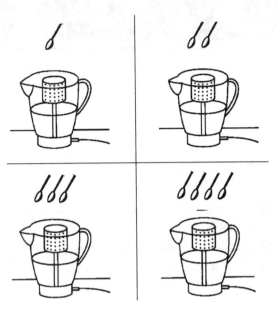

Please state the factors that were held constant and how they were kept the same.

_____   _____

_____   _____

_____   _____

_____   _____

SELF-CHECK ✔️

Constants	How they were kept the same.
Amount of water	same amount used each time
Type of coffee maker	identical coffee makers used
Length of brewing time	each brewed same amount of time
Type of coffee	same kind of coffee used in each pot

You may have listed other constants besides these. The manipulated variable in this experiment is the amount of coffee. If different amounts of water were used in each pot, for example, it would not be possible to say that the outcome was due only to the amount of coffee used.

What factors must be held constant in this suggested experiment?

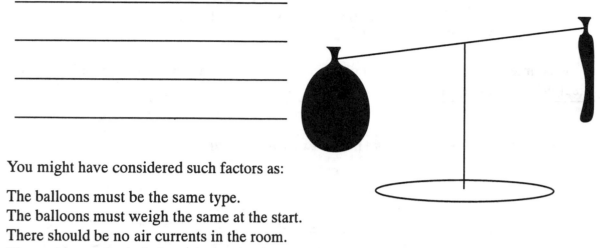

_____

_____

_____

_____

You might have considered such factors as:

The balloons must be the same type.
The balloons must weigh the same at the start.
There should be no air currents in the room.
The scale must be in balance before the balloons are attached.

Actually you can't be sure about these because you only have a picture to examine.

What factors are held constant in the following experiment?
A herd of Angora goats is divided into two groups. Both groups are housed in the same building, fed at the same time each day, and given the same amount of water. One group gets Brand X feed and the other Brand Y.

_____   _____

_____   _____

SELF-CHECK   ✔

        Type of goat   Feeding Time
        Type of housing   Amount of water

If you went beyond this list, you have listed constants that were not *explicitly* stated.

What additional factors are probably kept the same in the previous experiment, even though they are not mentioned in the description?

_____   _____

_____   _____

SELF-CHECK

<u>Number in each group</u>   <u>Age of the animals</u>
<u>Temperature of housing</u>   <u>Bedding characteristics</u>
<u>Amount of feed available</u>

This time you may have thought of others. This is all right if you believe that they could affect the responding variable.

What factors are kept the same in this experiment setup?

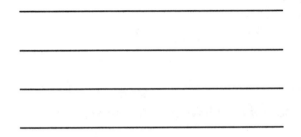 

SELF-CHECK

Amount of liquid   Kind of liquid
Size of container   Temperature of liquid
Shape of container   Pressure on liquid surface

By this time you are probably saying, *How in the world do they expect me to get all of those? Besides, I thought of some that they didn't.* This is just the point. In any experiment there could be many potential variables; so many that it is often difficult to think of them all. A good way to help you identify these potential variables is to make a list of the materials and environmental conditions used in an experiment. Then, think of ways to keep them the same.

Consider an experiment to test the prediction that the more light plants receive, the taller the plants will grow. Only two variables should be allowed to change or vary, amount of light (MV) and height of plant (RV). All other factors must be kept the same. The constants in this experiment include:

- All the plants are the *same* size
- The *same* soil in each pot
- Watered at the *same* time each day
- Given the *same* amount of water
- Kept in the *same* place

Suppose two groups of plants were used. One group of plants would receive more light than the other group of plants. All other potential variables would be kept the same for each group of plants. They would become the constants for this experiment.

By keeping all other factors the same, the experiment is a fair test of how amount of light affects the height of plants.

Suppose in this experiment one group of plants was grown at 16°C.

At what temperature would you keep the other group of plants? _____

SELF-CHECK ✓

    16°C
_____

What factors should be held constant to test the following hypothesis?
The greater the amount of salt added to ice, the lower the temperature of the mixture.

_____   _____

_____   _____

_____   _____

SELF-CHECK ✓

Several identical containers were filled with the *same amount* of ice. Use the *same kind of salt* in each and measure the temperature with the *same kind of thermometer* in each.

You may have thought of others. As long as you were attempting to keep everything in all setups the same except the amount of salt, it would be okay.
_____

So far in this program you have learned to identify constants. A constant is something, other than the manipulated variable that could affect the responding variable. In a controlled experiment the constants are prevented from affecting the outcome by making sure they are the same in all cases.

Scientists are interested in explaining events. To do so they conduct investigations to determine what effects manipulated variables have on responding variables. In order to plan what investigations should be conducted, a statement called a hypothesis is made. A hypothesis is a prediction about the effect a manipulated variable will have on a responding variable.

What two variables will you find included in a hypothesis?

_____   _____

SELF-CHECK ✓

    Manipulated variable   Responding variable

We could define a hypothesis as an attempt to predict an outcome.
Which of the following are stated as hypotheses?

☐   1. As more salt is dissolved in water, the water will become cloudy.
☐   2. The earth's crust contains 90 elements.
☐   3. Magnetism and gravity are not the same
☐   4. If the length of a vibrating string is increased, the sound will become louder.

SELF-CHECK ✓

1, 4

Remember, a hypothesis is stated as the predicted effect one variable will have on another.

Which of these statements are hypotheses?

☐   1. As the temperature of its environment increases, the temperature of a cold-blooded animal increases.
☐   2. Glass is harder than iron; therefore glass will scratch anything which is softer than iron.
☐   3. A change in weather causes a change in mood.

SELF-CHECK ✓

1,2,3

In all three cases we are predicting what will happen to a responding variable if we manipulate another variable.

Which of these is stated as a hypothesis?

☐   1. If clouds act as insulators, then the earth should get colder on cloudless nights.
☐   2. Leaves manufacture food, stems transfer food, and roots store the food in plants.

SELF-CHECK ✓

In #1 the effect of a manipulated variable (amount of cloud cover) on a responding variable (night time temperature) is predicted or hypothesized. *Therefore, #1 is a hypothesis.*

In #2, only results are reported. *Therefore, #2 is not a hypothesis.*

Which of these are hypotheses?

❑   1. The colder the temperature, the slower plants grow.
❑   2. The deeper one dives, the greater the pressure.
❑   3. Algae are living organisms.

SELF-CHECK ✔

1, 2

One could manipulate the temperature or the depth of the dive, but there is no variable to manipulate in #3.

Remember that a hypothesis states what effect a manipulated variable will have on a responding variable.

Now you try to write a hypothesis. Write a statement that predicts the outcome if the *amount of light* is one variable and the other is *plant growth*.

SELF-CHECK ✔

The greater the amount of light, the greater the amount of plant growth.
The less light the plant receives, the less the plant will grow.

These are just some of the hypotheses you could have written if you tried to predict what would happen to one variable as you manipulated another. Remember, you don't have to know for sure what the effect will be; you're only making a prediction, an educated guess. If you test your hypothesis, you should use only one manipulated variable. All other potential variables must be kept the same in your experiment. These factors become your constants.

You now have had some practice in identifying variables and hypotheses when given the parts of an investigation. Next you will analyze the entire investigation and identify the variables involved and the hypothesis being tested.

Here is the description of an investigation:

> John was interested in determining the effect the number of plants located in an area has on growth rate. He planted radish seeds in several milk cartons. In the first carton, he planted 5 seeds 1 cm deep and no less than 5 cm apart; in the second, 10 seeds were planted 1 cm deep and no more than 2 cm apart; in the third, 15 seeds 1 cm deep and 1 cm apart; and in the fourth, 20 seeds 1 cm deep and 0.5 cm apart. Each carton was watered daily and daily measurements of the length of leaves were made.

What were some of the constants?

_____   _____

_____   _____

What variable was manipulated? _____

Which variable was expected to respond? _____

What was the hypothesis being tested? _____

SELF-CHECK ✓

Some factors that were kept the same are: Kind of seed, planting depth, soil, environmental temperature, amount of water, kind of containers, and amount of light received. Since the number of plants in an area was manipulated and the length of the leaves was expected to respond, the hypothesis probably was: *As the number of plants in an area increases, the length of the leaves will become shorter.* Or it could have predicted that the leaves would be longer. In stating a hypothesis the decision as to what the effect will be can be based on data gathered in related situations or even a hunch or educated guess. Until data are gathered and interpreted, however, one prediction may be just as valid as another.

Here is a description of another experiment:

> Is there a relationship between the amount of training received and the length of time a learned behavior persists in insects? Select a number of sowbugs which always turn right when entering the intersection of a T-shaped maze. Using the tendency of sowbugs to avoid light, it is possible to train them to turn left by shining a strong-light from the right as they enter the intersection. Subject an animal to 1, 5, 10, 15, or 20 training sessions. Test each animal once an hour by running it through the T-maze.

What were some of the constants?

_____   _____

_____   _____

What variable was manipulated? _____

Which variable was expected to respond? _____

SELF-CHECK ✓

Type of Animal   Strength of Light Source
Shape of Maze   Environmental Temperature

These are just some of the constants. You may have stated others.

The manipulated variable was the amount of training, while the responding variable was the length of time a learned behavior persisted.

Sometimes it is possible to make inferences concerning variables and hypotheses given only the physical setup and the problem under study. Examine the stated problem and drawings below and answer the questions that follow.

Problem: Will different kinds of soil retain different amounts of water?

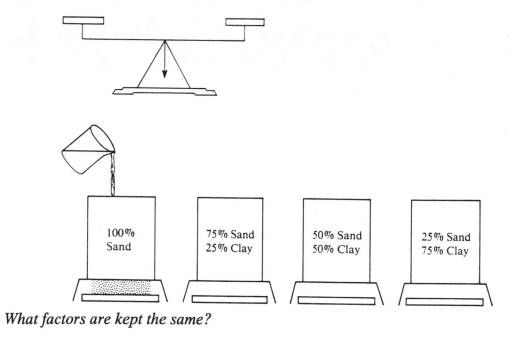

What factors are kept the same?

_____   _____

_____   _____

What variable is being manipulated? _____

Which variable is expected to respond? _____

What is the hypothesis being tested? _____

SELF-CHECK ✓

Factors that could vary, such as the amount of soil, size of container, soil temperature, kind of sand, and kind of clay, are kept the same. The percent of clay and sand is being manipulated and the weight of retained water is expected to respond. The hypothesis most likely being tested is: As the amount of sand in the soil decreases, the amount of water retained by the soil increases.

You now have had some practice in analyzing investigations, looking at the variables involved, and identifying the hypothesis being tested. In the next section you will begin the task of designing your own investigation. Among other things this requires you to construct hypotheses.

Now take the Self-assessment for Chapter 12.

Self-Assessment Analyzing Investigations

Read the description of this investigation and then answer the questions below.

1. A study was done to determine how the number of paper clips picked up was related to the number of dry cells connected to the electromagnet. The magnet, connected to 1, 2, 3, 4 or 5 D cells, is placed on the top of a pile of 100 paper clips and lifted.

    a. Identify the constants in the above investigation.

    _____   _____

    _____   _____

    b. Identify the manipulated variable.   _____

    c. Identify the responding variable.   _____

    d. State the hypothesis being tested.

    _____

2. Identify each of the statements below that is stated as a hypothesis.

    ❑   a. Baking powder is used in biscuits.
    ❑   b. The brighter the color of an orange, the juicier the fruit.
    ❑   c. Brass contains copper and zinc.
    ❑   d. The more antifreeze, the lower the freezing temperature.

3. Classify these statements as hypotheses (H) or non-hypotheses (N):

H N

❏ ❏   a. The more cabbage in the stew, the stronger the flavor.
❏ ❏   b. Most apples are red.
❏ ❏   c. The faster the river flows, the greater the erosion.
❏ ❏   d. Dental floss is waxed.

4. Suppose you wished to test the hypothesis stated below. Which of the factors listed should be kept the same in the experiment?
   Hypothesis: The warmer the water, the faster an aspirin will dissolve.

❏   a. amount of water
❏   b. brand of aspirin
❏   c. temperature of water
❏   d. size of the tablet

5. List four constants for an experiment if you were to test this hypothesis:
   Hypothesis: The thicker the coating of peanut butter, the slower a sandwich will be eaten.

SELF-ASSESSMENT ANSWERS

1.
   a. number of paper clips in the pile, type of dry cell, size of the electromagnet, shape of the pile.
   b. number of dry cells.
   c. number of paper clips picked up.
   d. the more dry cells connected to an electromagnet, the more paper clips it will pick up.

2.   b, d
3.   a. H   b. N   c. H   d. N
4.   a, b, d
5.   brand of peanut butter
     thickness of bread slice
     texture of the peanut butter
     color of the bread

Your answers may vary greatly. Consider them correct if they deal with the characteristics of the peanut butter and bread.

ASSESSING FOR SUCCESS: OPEN RESPONSE QUESTION[1]

Directions to the Student

Chris wants to find out which of two spot removers is better. First, he tried Spot-Remover A on T-shirts that had fruit stains and chocolate stains. Next, he tried Spot Remover B on jeans that had grass stains and rust stains. Then he compared the results.

a. What did Chris do wrong that would make it hard for him to decide which spot remover is better?
b. If you wanted to help Chris find out which spot remover is better, how would you design an experiment?

OPEN-RESPONSE 1

_____

_____

_____

_____

POINTS	SCORING GUIDE
4	Student recognizes kinds of material, stains, and spot removers as three variables to consider. Describes control of material and stain, while varying only spot remover. Clearly identifies a plan that will make comparison of the spot removers possible.
3	Student identifies both kinds of materials and stains as variables that must be held constant. May describe a design that eliminates variables, rather than ways of holding them constant. Misses some points of logic.
2	Students saw that Chris introduced variables in his design that were not held constant. One of those variables is identified, but a way of holding the variable constant may not be described, or it may be described incorrectly.
1	Student recognizes the problem of stain removal, but makes no comment on design. Does not recognize that variables have been introduced.
0	Blank

	EXAMPLES OF STUDENT RESPONSE* FOR EACH SCORING GUIDE LEVEL
4	a. He didn't try Spot remover A on the jeans and he didn't use spot remover B on the T-shirts. b. First I would have two T-shirts with fruit stains and chocolate stains on it. Then I would put stain remover A and stain remover B on it which ever got the stains out better would be the one he would use. I would have two pairs of jeans with grass and rust stains and I would put stain remover A on one of them and stain remover B on the other. What ever gets out the stains the best would be best.
3	a. He used two different kinds of clothing, and the clothing also had different kinds of stain on them. b. I would have the same kind of clothing and I would also have the same of stains on my clothing.
2	Chris used he spot removers on different stains. I would get spot remover a and spot remover b and use them on the same stains. The find my results.
1	I would do the same thing that he did but look a the close very carefully.
	*Student errors have **not** been corrected.

1. Source: Grade 4, Science Question 2, 1991-92 KIRIS Common Open-response item. Kentucky Department of Education, KIRIS Division, 500 Metro Street, Frankfort, KY, 40601

© Rezba, Sprague, Fiel, Funk, Okey, & Jaus. LEARNING AND ASSESSING SCIENCE PROCESS SKILLS, Kendall/Hunt 1995.

Sight
Smell
Sound
Taste
Touch

PURPOSE

An investigation or experiment usually begins with a problem that needs solving, a question that needs answering, or a decision that needs to be made. The integrated science process skills are problem solving and decision making tools used to gather information (data) and test inferences (explanations). We investigate to determine if cause and effect relationships exist between things. By deliberately changing one factor in an investigation, another may change as a result. Before any investigating or experimenting is conducted, a *hypothesis* is usually stated. Hypotheses are *predictions* about the relationships between variables. The hypothesis provides guidance to an investigation about what data to collect. In this chapter you will learn to write hypotheses and you will use this skill later when you plan and carry out your own investigations.

OBJECTIVES

After studying this chapter you should be able to construct a hypothesis when provided with a problem.

Approximate time for completion: 25 minutes

Begin by considering this problem:
What affects how fast a person can run a 100 meter dash?

Try to think of all the factors that might possibly affect how fast a person can run a 100 meter dash. There are factors related both to the individual and to the environment of the individual which could affect his speed. For example, lung capacity, muscle tone, length of legs, and motivation are characteristics of an individual which could affect his speed. The direction of the wind, surface of the track and type of shoes worn are characteristics of an individual's environment which also could affect his speed. If any of these factors could be changed, the outcome (the speed of the runner) might be affected.

Here is another problem for you to analyze:
How fast will an object fall through a liquid?

Identify variables that might affect the rate of a falling object. First, consider variables related to the object, then consider variables related to the environment of the object. (If materials are available in your supply area, try dropping different objects into different liquids)

What characteristics of the object might affect its speed as it falls through the liquid?

_____   _____

_____   _____

SELF-CHECK   ✓

> volume of the object   weight of the object
> shape of the object   density of the object

These are just a few. There are many others.

What characteristics of the environment of an object might affect its speed as it falls through the liquid?

List some possibilities.

_____   _____

_____   _____

SELF-CHECK ✓

Both the liquid and the container are part of the environment of the object so characteristics of either could have an effect.

Liquid
temperature of liquid
amount of liquid

Container
size of container
shape of container

Only some of the possibilities are listed. You may have come up with others.

Another problem for you to analyze is given below. You are to identify some variables which might affect the rate of growth of the plants. Remember to consider characteristics of both the plant and its environment.

What affects the rate at which a plant grows?

List some variables which might affect the rate of growth.

_____   _____

_____   _____

_____   _____

SELF-CHECK ✓

Plant
a. type of plant
b. age of plant

Light
a. amount of light
b. direction of light
c. color of light

Water
a. amount of water
b. type of dissolved
   minerals in water.

The list of possibilities is almost endless. Only a few are shown here.

Another problem for you to analyze is given below. This time identify some variables that might affect the exercise time.

Problem: What determines the amount of time an animal will spend in an exercise wheel?

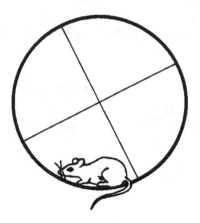

List some variables that might affect the exercise time.

Animal   **Environment of Animal**

_____   _____

_____   _____

_____   _____

SELF-CHECK ✓

Animal	**Food**	**Cage**
a. age of animal	a. amount of food	a. size of cage
b. sex of animal	b. feeding time	b. shape of cage
c. number of legs	c. type of food	

Once the variable of interest has been selected, a testable hypothesis can be stated. The term *testable hypothesis* is used because this indicates one of the functions a hypothesis should serve. A hypothesis should point the way towards the design of an investigation to test it. To construct a hypothesis, express what you think will be the effect of the variable you will deliberately change on the variable you expect to respond. This prediction can be based on fact, opinion, hunch, or whatever resources you may possess. For example, to construct a hypothesis related to the problem, *What affects the speed of a car?* one might select the variable *size of tires* to test. One could then construct a hypothesis in the following way: *As the size of its tires increases, the speed of a car decreases.*

In the following problem, construct a hypothesis for each variable selected for testing by predicting its effect on the hatching rate.

Problem: Russell raises bees. He noticed that different numbers of young hatched from the same number of hives at different times. He wondered what factors might influence the hatching rate of bees. He selected the following variables to be tested:

    a. temperature of the hive
    b. relative humidity inside the hive
    c. amount of food available
    d. number of bees living in the hive

Construct a hypothesis for each variable listed above.

    a. _____

       _____

    b. _____

       _____

    c. _____

       _____

    d. _____

SELF-CHECK ✓

    a. As the temperature of the hive *increases*, the hatching rate *will increase*.
    b. As the relative humidity inside the hive *increases*, the hatching rate *will decrease*.
    c. As the amount of food available *decreases*, the hatching rate *increases*.
    d. As the number of bees living in the hive *increases*, the hatching rate *decreases*.

Your prediction as to the effect of each variable is as good as ours. It will remain just that—a prediction, until someone tests it.

Problem: **What factors determine the rate at which an object falls through air?**

Possible variables:
a. volume of object
b. surface area of object
c. length of fall
d. weight of object

Construct a hypothesis for each variable.

a. _____
_____

b. _____
_____

c. _____
_____

d. _____

SELF-CHECK ✓

a. As the volume of an object *increases*, the rate at which it falls through air *decreases*.
b. As the surface area of an object *increases*, the rate at which it falls through air *decreases*.
c. The longer (or farther) an object falls through air, the faster it will fall.
d. The more weight an object has, the faster it will fall through air.

Your hypotheses may be entirely different from those above and still be correct. In each case, however, your statement should include the predicted effect of one variable on another variable.

Now, practice putting together the skills you have learned. You will be presented with several problem situations. Construct hypotheses for each problem.

List at least three possible variables and construct a hypothesis for each variable.

Problem: ***Why is it warmer in one house than another?***

    a. Variable 1: _____

       Hypothesis 1: _____

    b. Variable 2: _____

       Hypothesis 2: _____

    c. Variable 3: _____

       Hypothesis 3: _____

SELF-CHECK ✓

Outside temperature
The higher the outside temperature, the higher the temperature inside the house.

Location of house
The nearer the house is to the equator, the higher the temperature inside the house.

Slope of roof
The steeper the roof, the higher the temperature inside the house.

Thickness of insulation
The thicker the insulation, the higher the temperature inside the house.

Number of openings to the outside
The more openings (window and doors) to the outside, the lower the temperature inside the house.

These are only a few of the many possible hypotheses that could be constructed concerning this problem. Your hypotheses may be very different from these. However, each hypothesis must state how you think the manipulated variable will affect the responding variable.

Here is another problem. Generate a list of possible variables, choose three and construct a hypothesis for each related to the length of a shadow.

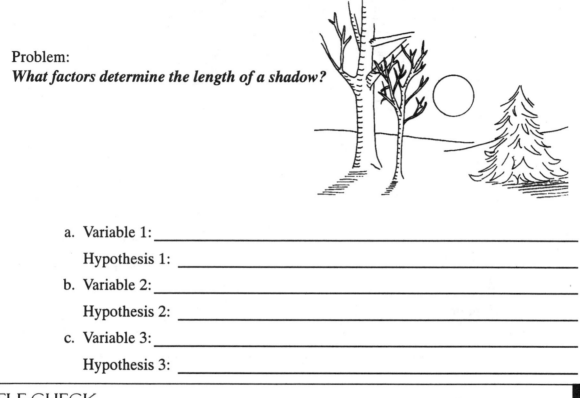

Problem:
What factors determine the length of a shadow?

    a. Variable 1: _____

       Hypothesis 1: _____

    b. Variable 2: _____

       Hypothesis 2: _____

    c. Variable 3: _____

       Hypothesis 3: _____

SELF-CHECK ✓

    a. *height of object*
       The taller an object, the longer its shadow.
    b. *time of day*
       The closer the time moves toward noon, the shorter the shadow of an object.
    c. *season of year*
       As the season progresses from summer to winter, the length of a shadow becomes longer.

You have now had an opportunity to construct several hypotheses. A second task required to design an investigation is deciding how you will measure the variables you have selected to test. This measurement problem will be discussed in the next chapter. Finally, in the last two chapters you will really *put it all together* and use skills learned in all these chapters to design and conduct some investigations. Now take the Self-assessment for Chapter 13.

Self Assessment | Constructing Hypothesis

For each of the following problems list three variables which could affect the responding variable. State a hypothesis for each variable listed.

1. Why doesn't an animal breathe at the same rate all the time?

    a. Variable 1: _____

    Hypothesis 1: _____

    _____

    b. Variable 2: _____

    Hypothesis 2: _____

    _____

    c. Variable 3: _____

    Hypothesis 3: _____

    _____

2. What determines how high a balloon will rise?

    a. Variable 1: _____

    Hypothesis 1: _____

    _____

    b. Variable 2: _____

    Hypothesis 2: _____

    _____

    c. Variable 3: _____

    Hypothesis 3: _____

    _____

228 • Chapter 13

SELF-ASSESSMENT ANSWERS

1. Here are a few of the many possible variables. Included with each is a possible hypothesis. Your answers may be different and still be correct.

   amount of exercise
   As the amount of exercise increases, the breathing rate will increase.

   age of the animal
   As the age of the animal increases, the breathing rate will decrease.

   temperature of environment
   As the temperature of the environment decreases, the breathing rate will increase.

   body size
   As an animal's body size increases, the breathing rate increases.

   altitude
   As the altitude increases, the animal's breathing rate increases.

2. *size of balloon*
   The larger a balloon is, the higher it will rise.

   weight of balloon
   The lighter a balloon is, the higher it will rise.

   temperature of air
   The cooler the air surrounding the balloon, the higher it will rise.

   temperature of balloon
   The warmer the balloon, the higher it will rise.

ASSESSING FOR SUCCESS: INTERVIEW OF INDIVIDUAL PERFORMANCE[1]

PREPARATION: When time and personnel resources allow, an individual interview can be a powerful assessment tool. The following example of an interview assessment was called an Individual Competency Measure and is from Formulating Hypotheses, a Science -A Process Approach module. Listed below are the objectives measured by this assessment task and the materials needed.

OBJECTIVES: At the end of this module the student should be able to

1. DISTINGUISH between: statements that are hypotheses and statements that are not.
2. DISTINGUISH between observations which support a stated hypothesis and those which do not.
3. CONSTRUCT a hypothesis from a set of observations.
4. CONSTRUCT a prediction based on a hypothesis.

MATERIALS

✓ Hypotheses and Predictions (Figure 12)
✓ Glass jars, 3, identical
✓ Water
✓ Steel rod
✓ Metric ruler

Directions to the Student

TASK 1 (Objective 1): Give the child a copy of Hypotheses and Predictions, Figure 12.

Say to the students: "I will read two statements on your sheet. Put an X before the statement which is an hypothesis." Read the following statements.

1. Thin liquids evaporate faster than thick liquids.
2. Alcohol evaporates faster than water.

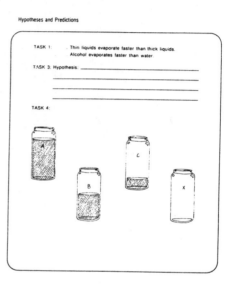

FIGURE 12

The child should mark the first statement.

TASK 2 (Objective 2): Say, "Assume that the hypothesis is that thin liquids evaporate faster than thick liquids. A scientist observed that water is much thinner than corn syrup and that water evaporates faster. Does this observation support the hypothesis?" **The child should reply that it does.**

TASK 3 (Objective 3): Put on a table in front of a child two identical glass jars, one of which contains a little water and the other of which is almost filled with water. Give the child a steel rod and say, "Tap both jars slightly on the sides and tell me about the sounds that are made. Be very specific when you make your observations." The child should say that the sounds are different and how they are different. If he does not say that the jar with more water has a lower pitch, ask, "Which jar has the lower pitch?" Put a third jar about half filled with water in front of the child. Say, "Tap this jar and tell me about the sound you hear. Then, order the jars from the tones of highest pitch to lowest pitch." When he has ordered the jars say, "Write on the data sheet a hypothesis about the water, the jars, and the sounds that are made when the jars are tapped with a steel rod." **The child should write something like this: When a glass jar is tapped with a rod a tone is heard. The greater the amount of liquid there is in the jar, the lower the pitch of the tone.**

TASK 4 (Objective 4): Identify the ordered jars by marking them with an A for lowest pitch, B for middle pitch, and C for the highest pitch. Say to the child, "Look at the drawings on the data sheet. Suppose I gave you a fourth jar X, that has a higher pitch than A but lower than B. Draw a line to represent the amount of water that would be in X." **The child should draw a line representing an amount of water in jar X that is between the amounts shown in jars A and B.**

---

1. Module 70, Formulating Hypotheses, Science - A Process Approach II. Copyright © 1986 by Delta Education, Hudson, NH. Used by permission.

Defining Variables Operationally

Sight
Smell
Sound
Taste
Touch

PURPOSE

During an investigation measurements of the variables are made. However, before making the measurements the investigator must decide how to measure each variable. In this chapter you will practice making decisions about how to measure variables.

OBJECTIVES

When you finish this chapter you should be able to :

1. State how the variables are operationally defined in an investigation when given a description of the investigation.
2. Construct operational definitions for variables.

Approximate time for completion: 25 minutes

By specifying a procedure for measuring a variable you are making an operational definition. To operationally define a variable means to decide how you will measure it. Thus an operational definition tells what is observed and how it is measured.

Different investigators may use different operational definitions for the same variable. For example, suppose an investigation was conducted to test the effects of Vitamin E on the *endurance of a person*. The variable *endurance of a person* could be defined many different ways:

        a. the number of hours a person could stay awake
        b. the distance a person could run without stopping
        c. the number of jumping jacks a person could do before tiring

Each of the above statements is an *operational definition* of the same variable.

Here is a brief description of an investigation. Your job is to determine how each of the variables was operationally defined in this investigation. That is, you are to say how the manipulated and responding variables in this investigation were measured.

INVESTIGATION

A study was done to determine if safety advertising had any effect on automobile accidents. Different numbers of billboards were put up in Cleveland over a period of four months to see if the number of people hospitalized because of auto accidents was affected. In March, five billboards carried safety messages; in April there were ten; in May there were fifteen; and in June there were twenty. During each of these four months, a record of the number of people hospitalized because of accidents was measured.

The variable *safety advertising* was manipulated in this investigation to see if *automobile accidents* would respond. How was each one operationally defined?

Safety advertising _____

_____

Automobile accidents _____

_____

SELF-CHECK

Safety advertising is operationally defined as the number of safety billboards put up in the city during each month. (Observed = safety billboards. Operation performed to measure what is observed = counting the number of billboards erected each month.)

Automobile accidents are operationally defined as the number of people who are hospitalized because of automobile accidents. (Observed = people who are hospitalized because of auto accidents. Operation performed to measure what is observed = counting the number of hospitalized people).

It is important to note that these variables, or any others, could be measured in a variety of ways. It is entirely up to the investigator how the variables in the study will be operationally defined. However, the operational definition should be explicit enough that another person could carry out the measurement without any further information from the investigator.

Now take a look at the following investigation. How were the manipulated and responding variables operationally defined in this study?

INVESTIGATION

A study was done to determine the effect that exercise has on pulse rate. High school students rode bikes for different numbers of kilometers and then their pulse rate was measured. One group rode 10 km, a second group rode 20 km, a third group rode 30 km, and a fourth group rode 40 km. Following the exercise the pulse rate was immediately measured by counting the pulse for one minute.

How was each variable operationally defined in this investigation?

MV: _____

_____

RV: _____

_____

SELF-CHECK ✓

Amount of exercise is the manipulated variable. It was measured by counting the number of kilometers a person rode. **Pulse rate** is the responding variable. It was measured by counting the number of heartbeats felt at the wrist following exercise.

Amount of exercise and pulse rate could have been operationally defined in other ways. For example, **amount of exercise** could have been defined by having the students run in place for designated periods of time. It could have been operationally defined in terms of the number of knee bends each student did. The point is that **amount of exercise** can be defined in most any way you want to in an experiment. There is a variety of ways you could define it. When you select one way to measure **amount of exercise** for your experiment, you have defined it operationally.

Here is a third investigation. Read it and write how the manipulated and responding variables were operationally defined in this study.

INVESTIGATION

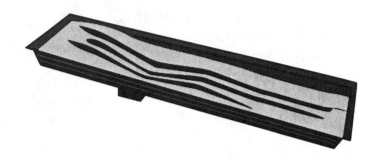

A study was conducted to see if the amount of erosion was affected by the slope of the land. The end of a stream table was raised to four different heights (10 cm, 20 cm, 30 cm, 40 cm) in order to make it slope different amounts. (A stream table is a plastic box about 40 cm wide and 100 cm long. Sand or soil is placed in the box and water can be run in at one end.) At each height a liter of water was poured in at one end of the stream table. After the water had run over the soil, the depth of the gully cut by the water was measured.

_____

_____

_____

_____

SELF-CHECK ✓

Amount of Erosion is the responding variable. It was operationally defined in this study as the depth of the gully cut by the water. *Slope of the land*, the manipulated variable, was operationally defined as the height to which the end of the stream table was raised

Clearly, these two variables could have been operationally defined in other ways. For example, *slope of the land* could be measured in degrees of tilt of the box. Measuring the mass of the eroded soil would be another way of operationally defining *amount of erosion*

To think of a variety of ways that a variable might be operationally defined, consider this case. Suppose you wanted to operationally define the variable *size of a person*. Write at least three ways this variable could be defined operationally.

_____

_____

_____

_____

SELF-CHECK ✓

Here are some ways that could be used to define the *size of a person*.

    a. the reading in kilograms obtained when a person steps on a scale.

    b. the smallest number showing when a person stands against a ruler which extends from the floor up.

    c. the amount of water that overflows when a person is submerged in a full bathtub.

    d. the amount of tape required to encircle the chest, waist and hips.

Now try another variable. Suppose you are an expert in agriculture and you are growing beans in an experiment. You need to operationally define the variable *amount of plant* growth. Write down three different ways that you could operationally define this variable. Just think of three different ways you could measure how much the plant grew.

_____

_____

_____

_____

SELF-CHECK ✓

Some possibilities you might have thought of are:

- Count the number of leaves on a plant. Wait two weeks and count them again.
- Measure the distance from the soil to the uppermost leaf. Ten days later, measure it again.
- Weigh the plant and its pot. Wait one month and do it again. The difference is how much it grew.

Suppose that an elementary school has a program underway for the purpose of increasing students' *enjoyment of reading*. What are some of the different ways that *enjoyment of reading* could be operationally defined? List at least three ways. Try to think of some specific things you could measure with your students in your classroom that would indicate their *enjoyment of reading*.

_____

_____

_____

_____

_____ ✓

SELF-CHECK

You might measure the variable *enjoyment of reading* in some of these ways:

        a. amount of time students voluntarily spend at the reading table
        b. number of references to books read during sharing time
        c. number of voluntary book reports
        d. number of books taken home

Some of these operational definitions may appeal to you more than others. If you were doing the experiment, of course, you could select the one you wished.

_____

You probably realize by now that most operational definitions are not found in the dictionary. Instead they have to be made up by the person who will use them. When you carry out an investigation, you have to decide how to measure the variables. You are constructing operational definitions when you make these decisions.

In the last part of this chapter you will be given three more variables to operationally define. For each you should try to think of a variety of ways to operationally define the variable.

Variable 1: Concern for Environment

Suppose that one of the goals of Springhill Elementary School is that all children acquire a *concern for their environment*. What are some of the ways that they might operationally define this variable? Describe at least three.

_____

_____

_____

_____

SELF-CHECK

Some possibilities are:

- the number of special projects students choose to do on environment matters
- the pounds of trash picked up on the playground each week
- the number of brown bags thrown away (instead of reused) from the lunch room
- the number of paper towels used in the washrooms
- the number of posters on environmental matters in a *show-your-concern-with-a-poster* contest

Obviously there would be many ways the variable ***concern for environment*** could be operationally defined. Very likely you had different and possibly better ideas than the list provided.

Variable 2: Understanding of Fractions

Suppose that you are a fifth grade teacher and you want your students to ***understand fractions***. What are some of the ways that you might operationally define this variable? Describe at least three.

_____

_____

_____

_____

SELF-CHECK

Some possibilities are:

In each case it is assumed that the students will be given the opportunity to demonstrate ***understanding***.

- Add fractions.
- Represent portions of a cut-up object with fractions.
- Select numbers that are fractions from a list.
- Reduce fractions to lowest terms.
- Solve real-life problems involving fractions.

These are a few of the ways that you might operationally define ***understanding of fractions*** for a fifth grader. In each of the above it is implied that there is some minimum amount of acceptable performance. There is room for much disagreement here. The operational definition one person chooses may be quite unacceptable to another and vice versa.

You may recognize the possibilities listed above as performance or behavioral objectives. They are statements of what students might be expected to do as evidence that they **understand fractions**. They are also operational definitions because they describe a procedure to use in measuring a variable.

Variable 3: Amount of Evaporation

An investigation is underway to see how the initial temperature of a liquid affects the **amount of evaporation**. Describe at least three ways that **amount of evaporation** could be operationally defined.

_____

_____

_____

_____

SELF-CHECK ✓

Three ways that you might have thought of are:

- a. Measure the depth of the liquid. Measure it again twenty-four hours later.
- b. Pour a known quantity of liquid into an open container. Measure its volume again three hours later.
- c. Mass the container of liquid. Twenty minutes later, mass it again. The difference is the amount of evaporation.

The following is a summary of what you should have learned in this chapter.

An *operational definition* describes how to measure a variable. It should state what operation will be performed and what observation will be made. Operational definitions have to be made up. There are usually various ways that one might choose to operationally define a variable. The definition you select depends on your intentions in an investigation.

Now take the Self-assessment for Chapter 14.

Self-Assessment Defining Variables Operationally

DEFINING VARIABLES OPERATIONALLY

1. Which of the following could be operational definitions for the variable *knowledge of trees*?

   ☐ a. identify at least fifteen different trees on a nature hike
   ☐ b. measure the average size of trees
   ☐ c. list at least twenty different trees that are native to your state
   ☐ d. match picture of trees with names on a test

2. How are the variables *amount of a liquid* and *solubility of salt* operationally defined in this investigation?

   An investigation is performed to see if the *amount of liquid* has any effect on the *solubility of salt* in it. (Solubility refers to the capability of dissolving a substance.) Four different amounts of water (50 mL, 100 mL, 150 mL, 200 mL) are placed in identical containers. Salt is added, five grams at a time, to each container. Each is stirred until no salt crystals can be observed in the liquid.

   Amount of liquid is _____

   _____

   Solubility of salt is _____

   _____

3. Describe three ways that you could operationally define the variable *size of automobile.*

   _____

   _____

   _____

SELF-ASSESSMENT ANSWERS

1. a, c and d could each be an operational definition.
2. *Amount of liquid* is measured in milliliters of water used.
   Solubility of salt is measured by mass of the salt dissolved.
3. Three possibilities are:

   a. Count the number of seats in the car.
   b. Measure the distance between the front and rear bumpers.
   c. Measure the distance between the two front tires.

Student Assessment **Example** ☆

ASSESSING FOR SUCCESS: OPEN-RESPONSE QUESTION

Preparation

One characteristic of the assessment reform movement of the 90's is the blurring of distinctions between instructional activities and assessment items. To use *Balloon Rockets* as an assessment task, provide students with a copy of Activity 2, The Balloon Rocket, found in Chapter 11 on page 197, and a copy of the following directions.

Directions to the Student

Use *The Four Question Strategy*[1], like you used in class on other topics, to brainstorm a list of variables on the question, *What affects the thrust of a balloon rocket?*. Questions 1 and 2 are already completed for you. You complete Questions 3 and 4.

1. What materials are readily available for conducting experiments on (balloon rockets)?
    Response: balloons, straws, string
2. How do (balloon rockets) act?
    Response: balloon rockets fly
3. How can you change the set of (balloon rocket) materials to affect the action?

4. How can you measure or describe the response of (balloon rockets) to the change?

Scoring

Students' responses should include at least two ways each of the materials in Question 3 can be changed or varied, and in Question 4 at least two ways *thrust* can be measured (operationally defined). Acceptable answers would include:

3. How can you change the set of (balloon rocket) materials to affect the action?

Response:	**Balloons**	**Straws**	**String**
	diameter	brand	length
	length	type of material	type of angle
	type of material	diameter	tightness
	# of puffs	length	type of material
	shape	mass	smoothness

4. How can you measure or describe the response of (balloon rockets) to the change?
    Response: Measure the time the rocket takes to cross the room.
    Measure the distance the rocket travels along the string.
    Time how long the rocket is moving.
    Figure out how fast the rocket travels in meters per second.

1. Cothron, J., Giese, R., and Rezba, R. (1993). Students and Research: *Practical Strategies for Science Classrooms and Competitions*. Dubuque, Iowa: Kendall/Hunt Publishing Company.

Designing Investigations

PURPOSE

In this chapter you will practice designing investigations to test hypotheses. Your skill in designing investigations will be limited only by your imagination. However, this does not mean that your design must be complicated. Quite the contrary, the simpler the design the more likely you will be able to collect usable data.

OBJECTIVES

When you have finished this chapter you should be able to design an investigation to test a given hypothesis.

Approximate time for completion: 30 minutes

An investigation can be defined as the setting up of a planned situation; the situation is planned to yield data that will either support or not support your hypothesis. If the manner in which a variable can be manipulated and the type of response expected is clearly stated in the hypothesis, then much of the work in planning how to collect data has been done. There remains the task of specifying conditions under which the work will be carried out.

Suppose you want to test this hypothesis:

Hypothesis

> ***The greater the surface area of a liquid exposed to the air, the faster evaporation will occur.***

The following investigation could be designed.

Design

Pour 100 mL of water at room temperature into each of five aluminum pans that are 5,6,7,8 and 9 cm square. Leave the pans sitting in an open room. After two hours have passed, measure the volume of water remaining in each.

Notice that the design consists of operationally defining both the manipulated and responding variables:

MV - leaving the liquid in different size open containers
RV - measuring the volume of the liquid before and after a specific time.

The design also states which other factors will be held constant. Factors that are kept the same in an experiment are called *constants*.

Although most of the investigations used for illustrations in this book have used five or six values for the manipulated variable, this number is by no means sacred. The number of values to be investigated depends entirely upon the investigator. It is important, however, to gather enough data to test the hypothesis and establish the relationships between the variables.

The investigator must decide how many different values of the manipulated variables are appropriate and how they should be selected. For example, if investigators were interested in the effect of temperature in a plant-growing experiment, they would probably not select 10°C, 12°C, 14°C, 16°C, and 18°C as the temperatures in which to grow plants. Instead, they would probably select values all the way from boiling to freezing in order to measure temperature effects on plants growing in a wide variety of conditions.

Read the following example of a hypothesis and the design of an investigation to test it. Then answer the questions about the design.

Hypothesis

The farther a ball drops, the higher it will bounce.

Design

Release a ball 10 cm above a rigid surface. Record the biggest number that appears under the ball as it rises beside the measuring stick. Repeat this procedure by dropping the ball from heights of 20, 30, 40, and 50 cm .

Questions

a. How was the manipulated variable operationally defined?

_____

b. How was the responding variable operationally defined?

_____

c. What were the constants?

_____

d. What values of the manipulated variable were selected for the investigation?

_____

SELF-CHECK ✓

    a. The height of a ball above a surface before it was dropped.
    b. Biggest number that appears under the ball as it rises.
    c. The same ball is dropped each time on the same surface. The same measuring stick would probably be used although this is not necessary if the sticks are equivalent. The same environmental conditions should prevail throughout the investigation (e.g., amount of wind, temperature, humidity, and so on).
    d. 10, 20, 30, 40, and 50 cm.

The questions you just answered should be considered *each* time you design an investigation. The design of an investigation should include:

1. a description of how the manipulated variable is operationally defined
2. a description of how the responding variable is operationally defined
3. a description of what factors are kept constant
4. the values of the manipulated variable selected for the investigation

Below is another example of a hypothesis and the design of an investigation to test it. Read the material and answer the questions about the design.

Hypothesis

The greater the concentration of carbon dioxide in the atmosphere, the greater the breathing rate of an animal.

Design

Using a breathing mask, feed air containing 0.01, 0.02, 0.03, and 0.04% carbon dioxide to different guinea pigs. Count the number of chest movements associated with breathing for five one-minute periods and average for each animal. Use similar sized animals of the same sex and general health.

Questions

    a. How was the manipulated variable operationally defined?

    _____

    b. How was the responding variable operationally defined?

    _____

    c. What were the constants?

    _____

    d. What values of the manipulated variable were selected for the investigation?

    _____

SELF-CHECK ✓

     a. The percentage of carbon dioxide in the air breathed by the guinea pigs.

     b. The average number of chest movements during five one-minute periods.

     c. Size, sex, general health, and species of animal. One can also infer that the general environment of the animals would be kept the same throughout the experiment.

     d. 0.01, 0.02, 0.03, and 0.04% carbon dioxide in the air.

Now that you have learned to identify the four parts of a design, try designing an investigation yourself. Design an investigation to test the following hypothesis. Be sure to include how the variables are to be operationally defined, how other factors will be held constant, and what values of the manipulated variable will be used. All of this is usually written in a brief paragraph.

Hypothesis

> ***The greater the amount of protein in the food of an animal, the more rapid its rate of growth.***

Design

_____

_____

_____

_____

_____

_____

_____

SELF-CHECK ✓

Here is just one of several possible designs.

Select a group of five newly weaned guinea pigs of the same size and sex. Feed each animal a basic diet of cereal pellets. Add daily to the diet of one animal 5 g of protein supplement, add 25 g daily to another, 50 g to the third, and 100 g to the fourth. Record the weight of each animal weekly for five months.

Another hypothesis is given below. Design an investigation to test this hypothesis. Remember to include each of the four parts described previously in your design.

Hypothesis

The greater the concentration of soap in a soap-water mixture, the larger the soap bubble that can be blown.

Design

_____

_____

_____

_____

_____

_____

_____

SELF-CHECK ✓

Here is one possible design.

Dip the bowl of a child's bubble pipe into a mixture of one part liquid soap and twenty parts water. Blow ten bubbles and break each on a clean sheet of white paper. Measure the diameter of the splatters and find the average splatter size. Repeat the procedure for a mixture of one part soap and 15, 10, 5 and 1 part water.

Now take the Self-assessment for Chapter 15.

Self Assessment Designing Investigations

DESIGNING INVESTIGATIONS

Design an investigation to test each of the following hypotheses:

1. Equal changes in the force needed to stretch a rubber band will cause equal changes in the length of the rubber band.

_____

_____

_____

_____

_____

_____

2. As the amount of water a plant receives each day increases, the amount of plant growth increases.

_____

_____

_____

_____

_____

_____

SELF ASSESSMENT ANSWERS

These are just two of many possible designs.

1. Suspend the rubber band from some point. Using a hook shaped from a paper clip, hang a washer on the rubber band. Measure the distance from where the rubber band is suspended to the paper clip hook. Continue adding washers to the hook one at a time until a total of ten washers are hanging from the rubber band. After each addition, measure the distance between suspension and hook. Using the same rubber band during all measurements will prevent other variables from affecting the outcome. Repeat the procedure two more times for a total of three trials.
2. Obtain 25 bean seedlings each planted in similar pots with the same soil. Place on a window ledge in indirect light. Each day give 5 plants 10 mL of water, 5 plants 20 mL, 5 plants 30 mL, 5 plants 40 mL, and 5 plants 50 mL of water. Every second day measure the distance from the soil to the uppermost point on each plant.

RATING SHEET - INDIVIDUAL PERFORMANCE WITHIN A GROUP

Directions to the Student

For each question, check the box that describes your behavior in the group during this task. Ask each person in your group to also rate your group behavior.

Student Name _____   Task Title _____   Date _____

Check one box per question.

	Almost Always	Often	Some-times	Rarely
Group Participation				
1. Joined in group discussion.				
2. Did his or her fair share of the work.				
Staying on Topic				
3. Paid attention, listened.				
4. Helped group get back to topic.				
Offering Ideas				
5. Suggested helpful ideas.				
6. Offered helpful comments on the ideas of others.				
Consideration				
7. Made good comments on the ideas of others.				
8. Gave credit to others for their ideas.				
Involved Others				
9. Helped others to join in by asking questions or asking for more information or ideas.				
10. Helped group members to reach group agreement.				
Communicating				
11. Spoke clearly. Was easy to hear.				
12. Explained ideas so they could be understood.				

Adapted from materials developed by the Connecticut Common Core of Learning Assessment Project, Connecticut State Department of Education, Bureau of Evaluation and Student Assessment, Hartford.

Experimenting

Sight
Smell
Sound
Taste
Touch

PURPOSE

Experimenting is the activity that puts together all of the science process skills you have learned previously. An experiment may begin as a question. From there the steps in answering the question may include identifying variables, formulating hypotheses, identifying factors to be held constant, making operational definitions, designing an investigation, conducting repeated trials, collecting data, and interpreting data. You will be expected to do all of these in this chapter as you plan and conduct an investigation of your own.

OBJECTIVE

Following the completion of this chapter, you should be able to construct a hypothesis, design and conduct an investigation for a problem you have identified or chosen to study.

Approximate time for completion: One hour

Now is the time for you to apply what you have learned in the earlier chapters. Several problems will be suggested from which you may select one to study. The problems are stated as questions that you or your students might ask. You are well equipped to find the answers to them.

If none of the questions on the following list appeal to you, ask and answer one of your own. Perhaps you are teaching a science unit that has given rise to questions like the ones below. Use one of those questions if you wish.

These are some of the problems from which you can select. Read them over and then follow the directions on the next page.

1. What affects the amount of time it takes a seed to sprout?
2. What affects the rate at which a person breathes?
3. What affects the amount of salt that can be dissolved in water?
4. What affects the time it takes water to freeze when placed in the freezer section of the refrigerator?

251

5. What affects the amount of gas produced when vinegar and baking soda are mixed together?
6. What affects how far a rubber band will fly?
7. What affects how far a bath towel can be pulled down across the towel rack before it begins to slide off?
8. What affects the rate at which an object falls through a liquid?
9. What affects the size of a bead of water?
10. What affects the rate at which water flows out of a bottle?

To investigate any of these questions, you should carry out and report the following. The report need not be elaborate. One piece of paper should be sufficient to contain all the information in your report. It should include:

1. the statement of the question or problem you are investigating
2. the statement of the hypothesis you are testing
3. a written description of the design of the investigation you will use to test the hypothesis (Remember to describe how the variables are operationally defined, factors to be held constant, and what values of the manipulated variable you will use.)
4. reporting the data in a table including repeated trials
5. constructing a graph of the data
6. a statement of the relationship observed between the variables
7. a comparison of your findings with your initial hypothesis to see if the hypothesis was supported or refuted by your investigation.

If you have questions and would like to see an example of a completed experiment, turn to the next page. Our investigation of "What affects how fast salt dissolves in water?" is given as an example.

When you finish this investigation you will have completed your study of the science process skills. How will you help students in acquiring these skills?

SAMPLE INVESTIGATION

1. *Problem:* What affects how fast salt dissolves in water?
2. *Hypothesis:* The greater the quantity of salt, the longer it will take to dissolve.
3. *Design:* Differing amounts of salt (6, 12, 19, 24, and 30 grams) will be measured and placed in 250 milliliters of water. The water will be stirred until no more salt crystals are observed and the length of time it takes the salt to disappear will be recorded. The procedure will be repeated two more times and the average time to dissolve will be calculated. The constants are: temperature of the water, kind of salt used, and the manner of stirring.
4. *Data Table:*

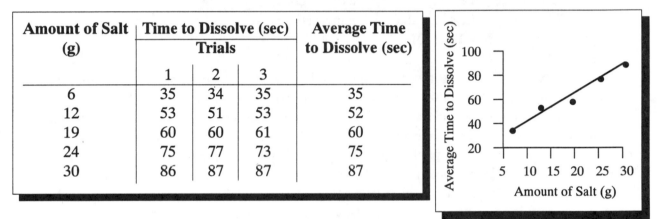

Amount of Salt (g)	Time to Dissolve (sec) Trials			Average Time to Dissolve (sec)
	1	2	3	
6	35	34	35	35
12	53	51	53	52
19	60	60	61	60
24	75	77	73	75
30	86	87	87	87

How Does the Amount of Salt in Water Affect Dissolving Time?

6. *Relationship Observed Between the Variables:* The greater the amount of salt added to the water, the longer it took to dissolve.
7. *Findings:* The data supported my hypothesis.

ADDITIONAL INVESTIGATIONS FOR INTEGRATED PROCESS SKILLS

Anything from television commercials to observations of natural phenomena can pose interesting problems that can be studied using the integrated process skills. Some ideas for investigations or projects you may wish to do with your own students follow.

1. What affects the absorbency of tissues or paper towels?
2. What affects the results of soft drink taste tests? (For example, would the temperature of the soft drink make a difference?)
3. Is there a relationship between the length of a person's arm and its power?
4. What affects how big a balloon can be inflated using one's lungs?
5. What affects one's reaction time?
6. What affects how quickly one tires?
7. What affects how high a ball will bounce?
8. How much water does a plant need?
9. Can a plant get too much fertilizer?
10. Is there a relationship between the size of a seed and its germination time?

11. What affects the rate at which heat passes through a solid?
12. What affects the size of an inflated balloon?
13. Do some leaves grow faster than others?
14. Does the color of food affect our choice?
15. What affects the boiling point of a liquid?
16. What affects the freezing point of a liquid?
17. What determines the effectiveness of a detergent?
18. What affects the absorbency of paper towels?
19. What affects the dissolving rate of an aspirin?
20. What affects the *fastness* of a fabric dye?
21. What affects the strength of concrete?
22. What affects the strength of a single strand of hair?
23. What affects the strength of a thread?
24. What affects the brightness of a burning candle?
25. What affects the length of service obtained from a flash light battery?
26. What affects the strength of a commercial adhesive?
27. How waterproof are paints?

It should be noted that in some investigations the values of the manipulated variables are distinct types or discrete categories, such as brands of paper towels, types of wood, gender, days of the week, and kinds of batteries. Discrete means that the categories are separate and not continuous. The spaces or intervals between categories are not equal and thus, have no meaning; there is no brand of paper towel that is halfway between Brand X and Brand Y. When the manipulated variable consists of discrete categories, such as brands of paper towels - Brand X, Brand Y, and Brand Z, you must display the data as a **bar graph**, not a line graph.

A **line graph** is appropriate whenever the variables are continuous. Continuous means that the values of a variable are not separate categories and that the intervals between them have meaning. Look at example #18. What affects the absorbency of paper towels? Suppose you use different amounts of paper towels, 1, 2, 3, 4 and 5 towels to see how number of towels affects the mass of water absorbed. Because number of towels is a continuous variable, you could also try 2 ¹/₂ sheets. The interval between 2 and 3 sheets has meaning. Other examples of continuous variables are volume of water, heights of ladders, units of clock time, and mass of fruit produced.

Does the Brand of Towel Affect How Much Water Is Absorbed?

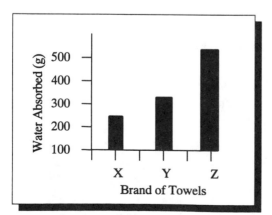

Does the Number of Towels Affect How Much Water Is Absorbed?

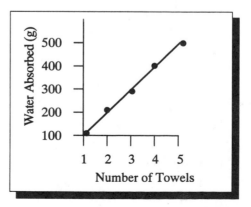

ASSESSING FOR SUCCESS: PERFORMANCE TASK

The Integrated Assessment System - Science[1], is an example of a commercially available science performance assessment. Tasks have been designed to use with students at each grade level in grades 2-8. For grades 3-8 tasks, students write a question that they can answer using the materials provided, such as *What will happen if I...?* Students are asked to describe how they performed their experiment, to draw pictures or make charts, to record observations, to measure, and to describe things that did not work the way they expected. A grid is provided for drawing pictures or constructing a graph. They also answer six questions about the experiment, describe how they would improve the experiment, and tell what they learned.

In one of the tasks for third grade, students are asked to design new swings for their school. Using paper cups, paper clips, weights, and string, students investigate which type of swing will swing best and what makes swings go faster or swing longer.

Scoring

Each *Science Performance Assessment* task can be scored two ways. Holistic scoring offers an overall judgement about the quality of problem-solving skills displayed; analytic scoring looks at specific aspects of problem solving. Analytic scoring evaluates specific skills such as experimenting, collecting data, drawing conclusions, and communicating.

Scoring services are available from The Performance Assessment Scoring Center of The Psychological Corporation by trained scorers. School districts may also score their own assessments using a comprehensive Scoring Guide.

---

1. From Integrated Assessment System - Science Performance Assessment Student Booklet. © 1992, The Psychological Corporation, Harcourt Brace Jovanovich, Inc. P.O. Box 839954, San Antonio, TX 78283-3954, (800) 228-0752. Used with permission.

DECISION MAKING 2

Now that you have learned the Integrated Science Process Skills, you can use what you have learned to improve existing science curricula. In learning the science process skills, you not only mastered the skills, but you also learned something about how these skills can be learned. By using this knowledge you can begin making some important instructional decisions about teaching science, especially the science process skills. In this section you will focus on the *application* of what you know about the integrated science process skills to improve elementary and middle school science textbook activities. The decisions you make can significantly enhance the quality of science in which your students are engaged.

Read *Textbook Activity Example C* on the next page. Think about how you might change the activity to better emphasize the science process skills.

As you study the sample activity, look at both the content and skills your students will be learning and how they would be learning them.

Ask yourself, *How will I provide opportunities for my students to:*

- *formulate or choose questions to study?*
- *state hypotheses?*
- *identify and define variables operationally?*
- *design investigations?*
- *conduct investigations and acquire data?*
- *construct tables of data?*
- *construct graphs?*
- *describe relationships observed between variables?*
- *describe the findings of an experiment?*

With these questions in mind, write what you consider to be strengths and weaknesses of *Example C* on a separate piece of paper. Then consider how you might change the activity to improve the weak areas.

After you have studied *Example C* and thought about how you might change it, turn the page and look at the annotated version of *Example C*. The changes made to *Example C* are only a few modifications that could be made to this activity to better emphasize the process skills. Your ideas for modifying this activity may have been different and even better.

257

Textbook Activity Example C

Can Work Be Measured?

You Will Need:

- ✓ 1 ruler with a groove running end to end
- ✓ 1 marble
- ✓ 1 small block of sponge, (about 2 cm x 3 cm x 6 cm)
- ✓ 1 other ruler, metric ruled

Follow This Procedure:

1. Place the grooved ruler on the table in front of you.
2. Place the sponge block at one end of a ruler so that the sponge touches the end of the ruler.
3. Raise the opposite end of the ruler 1 cm above the table.
4. Place the marble at the top of the ruler groove and let it roll into the sponge block.
5. Measure how far the sponge moved. Record your results in the table. The greater the distance the sponge moved, the more work the marble did on the sponge.
6. Repeat this activity raising the height of the marble to 3 cm, then 6 cm and 9 cm above the top of the table. Record your results.

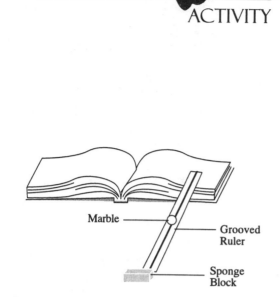

Marble — Grooved Ruler — Sponge Block

Record Your Results

Height of Marble	Distance the Sponge Moved
1 cm	
3 cm	
6 cm	
9 cm	

What Did You Find Out?

1. At what height did the marble have the most potential (stored) energy?
2. How was the amount of work done affected by the marble's potential energy?

Here are our suggestions. There are also many other ways to emphasize the integrated science process skills in this activity, such as asking students to identify the constants in this investigation or asking them to formulate a hypothesis and then state whether the data supported or did not support their hypothesis.

Textbook Activity Example C

Can Work Be Measured?

Part 1 modify activity by adding repeated trials. Also add column for average.

You Will Need:

- ✓ 1 ruler with a groove running end to end
- ✓ 1 marble
- ✓ 1 small block of sponge, (about 2 cm x 3 cm x 6 cm)
- ✓ 1 other ruler, metric ruled

Follow This Procedure:

1. Place the grooved ruler on the table in front of you.
2. Place the sponge block at one end of a ruler so that the sponge touches the end of the ruler.
3. Raise the opposite end of the ruler 1 cm above the table.
4. Place the marble at the top of the ruler groove and let it roll into the sponge block.
5. Measure how far the sponge moved. Record your results in the table. The greater the distance the sponge moved, the more work the marble did on the sponge.
6. Repeat this activity raising the height of the marble to 3 cm, then 6 cm and 9 cm above the top of the table. Record your results.

Marble — Grooved Ruler

Sponge Block

Record Your Results *Put on board - add repeated trials*

Height of Marble (cm)	Distance the Sponge Moved (cm)			Average distance moved
	1	2	3	
1 cm				
3 cm				
6 cm				
9 cm				

Part 2 (In groups of 4) In post lab discussions ask students to choose some other variable they might manipulate (size of marble, mass of marble, other rolling object, length of ruler runway, roughness of surface, and so on). Using their chosen IV have students design & conduct own investigation. Make table of data & graph. Report results to class.

What Did You Find Out?

1. At what height did the marble have the most potential (stored) energy?
2. How was the amount of work done affected by the marble's potential energy?

Here is another textbook activity example. Your task is to modify this activity to emphasize the process skills as modeled in the previous example. It may help you to review the questions on page 257 and to describe this activity's strengths and weaknesses. Then make your changes right on this activity page. When you are done, see the next page for modifications we made.

Textbook Activity Example D

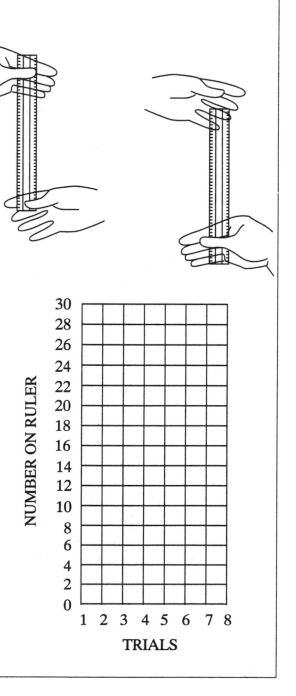

Measuring Your Reaction Time

You Will Need:

✓ a partner
✓ 1 metric ruler

Follow This Procedure:

1. Stand or sit facing your partner.
2. Hold your index finger and thumb open while your partner suspends the end of a ruler between them. Your fingers should be at the 0 cm mark.
3. Watch the ruler closely. When your partner drops the ruler, catch it between your fingers. Record the number where your fingers caught the ruler.
4. Repeat steps 2 and 3 seven more times.
5. Graph the data you collected for all eight trials.
6. Exchange places with your partner and repeat steps 2-5.

Write Your Conclusions:

How did the number of trials affect your reaction time?

Here are some modifications to this activity that <u>we</u> made. Your modifications may be even better.

Textbook Activity Example D

Use this activity to focus on stating hypotheses and setting up the design of an investigation used to test the hypothesis. Would also use to assess students' ability to state hypotheses and design experiments.

Measuring Your Reaction Time

You Will Need:

✓ a partner
✓ 1 metric ruler

Follow This Procedure:

1. Stand or sit facing your partner.
2. Hold your index finger and thumb open while your partner suspends the end of a ruler between them. Your fingers should be at the 0 cm mark.
3. Watch the ruler closely. When your partner drops the ruler, catch it between your fingers. Record the number where your fingers caught the ruler.
4. Repeat steps 2 and 3 seven more times.
5. *Make a table for the data collected.*
6. ~~5.~~ Graph the data you collected for all eight trials. *Draw a line of best-fit.*
7. ~~6.~~ Exchange places with your partner and repeat steps 2-5.
8. *State the hypothesis being tested by this experiment.*

Write Your Conclusions:

How did the number of trials affect your reaction time?

9. *State another hypothesis that could be tested using this activity (comparing reaction times by gender, age, time of day, and so on.)*
10. *Describe the design of an investigation that would test the hypothesis.*

(graph with vertical axis labeled NUMBER ON RULER, marked 0, 2, 4, 6, 8, 10, 12, 14, 16, 18, 20, 22, 24, 26, 28, 30; horizontal axis labeled TRIALS, marked 1 2 3 4 5 6 7 8)

MODIFYING REAL TEXTBOOK ACTIVITIES

The textbook examples you have just studied are typical of the kinds of science activities you might find in an upper elementary or middle school textbook. To gain a little more experience at modifying materials to emphasize the science process skills and to become acquainted with real textbook activities, you have one more task to complete.

Obtain either an elementary science textbook for grades 4, 5, or 6, or a middle school science textbook. Locate an activity and modify it to better emphasize the integrated science process skills just as you did in the previous examples. You might refer again to the questions on page 257 that focus on these skills. You may find it helpful to think of this task as a three-step procedure:

1. Examine the activity.
2. Identify parts that could be improved.
3. Improve it.

For feedback on your modifications, see your instructor, or try your modified activity with children and assess their skills.

At this point you have spent considerable time and effort in learning the science process skills. You have also been asked to think about how you will teach these skills to children and how you might assess how well students have learned these skills. How has all of this impacted your perception of your role as an elementary or middle school teacher of science?

Turn the page and complete the section called **Growing Professionally**.

Now that you have completed the activities and readings in *Learning and Assessing Science Process Skills*, return to the goal setting exercise on page xviii.

Read the statements you wrote describing the achievements for which you would like to be known after teaching science for three years. If your earlier goals have changed, rewrite your statements to reflect these changes.

Finally, in the space below, write two more things you will do to continue your professional development toward becoming an effective teacher of science.

APPENDIX

Materials and Equipment
for Learning and Assessing
Science Process Skills

EQUIPMENT AND MATERIALS

The materials listed below are needed to do the exercises in this book. Many of the materials can be found at home or purchased locally. The most expensive item on the list is a double-pan balance and masses. No materials are listed for Chapter 16 because the materials depend on which problem is chosen to be investigated.

Item	Amount	Used in Chapters
plant	1	1
sugar cube	1 box	1
birthday candle	box	1
clay	25 grams	1
matches	1 book	1
effervescent material	1 box	1
clear plastic drinking cups	1 package	1
plastic spoons	1 box	1, 7, 11
balances and masses	1	1, 4
meter stick and metric ruler	1	1, 4, 5
sensory materials (variety of objects to smell, taste, feel, hear and see)	1 set	2
magnetic compass (optional)	1	2
tangrams (page 269)	1 set	2
scissors	1	2
numbered buttons labeled 1-6	2 sets of 6	3
cereal box information panels	an assortment	3
peanuts in the shell	1 bag	3
pasta shapes	an assortment	3
centicubes	20	4
baby food jars (140 mL)	5	4, 7
large nonlead or coated sinkers (28 g size)	10	4, 5

liter containers	1	4, 7
assorted containers (various sizes and shapes)	4	4. 7
graduated cylinder 50 mL or larger	1	4, 11, 7
marbles	5	4
washers	4	4
ice cubes	as needed	4
thermometer (0-120°C)	1	4, 7, 11
toothpick	1 box	4
coin	1	5
pencil (new, sharp)	1	5
plastic sandwich bags	1 box	5
magnetic compasses	2	5
battery, size C or D	1	5
copper wire, 50 cm length, insulated, ends stripped	1	5
magnet	1	5
mystery box	1	5
string or cord	1 ball	6, 11
buttons or beads	25 red	6
	10 blue	
	10 green	
	5 white	
calcium chloride	500 grams	7
safety goggles	several	7
identical containers (baby food or plastic cups)	4	7
large container (for holding at least 1 liter of water)	1	7
sugar	container	11
100 mL pyrex beakers	4	11
measuring spoon	1	11
masking tape	1 roll	11
balloons	1 bag	11
plastic straws	1 box	11
quart size sealable bags	1 box	11
hot plate	1	11
timer	1	11
rubber tubing or other flexible hose	4 tubes (about 50 cm long, each with a different diameter)	11

Note: 1. With equipment such as balances, graduated cylinders, and plants, it would be better to have two to four of each for larger classes.
   2. The optional activity in Chapter 6 requires a jar and enough particles (marbles, peas, rice, and so on) to fill the jar.